Lecture Notes in Earth Sciences 85

Editors:
S. Bhattacharji, Brooklyn
G. M. Friedman, Brooklyn and Troy
H. J. Neugebauer, Bonn
A. Seilacher, Tuebingen and Yale

Springer
Berlin
Heidelberg
New York
Barcelona
Hong Kong
London
Milan
Paris
Singapore
Tokyo

Christoph Clauser

Thermal Signatures of Heat Transfer Processes in the Earth's Crust

With 9 Tables and 67 Figures, 25 in colour

Springer

Author

Dr. Christoph Clauser
Joint Geophysical Research Institute (GGA)
Stilleweg 2, D-30655 Hannover, Germany
e-mail: c.clauser@gga-hannover.de

English Translation

Dr. Clark Newcomb
Federal Institute of Geosciences and Mineral Resources (BGR)
Stilleweg 2, D-30655 Hannover, Germany

Cataloging-in-Publication data applied for

Die Deutsche Bibliothek - CIP-Einheitsaufnahme

Clauser, Christoph:
Thermal signatures of heat transfer processes in the Earth's crust / Christoph Clauser. - Berlin ; Heidelberg ; New York ; Barcelona ; Hong Kong ; London ; Milan ; Paris ; Singapore ; Tokyo : Springer, 1999
(Lecture notes in earth sciences ; 85)
ISBN 3-540-65604-9

"For all Lecture Notes in Earth Sciences published till now please see final pages of the book"

ISSN 0930-0317
ISBN 3-540-65604-9 Springer-Verlag Berlin Heidelberg New York

Typesetting: Camera ready by author
SPIN: 10694966 32/3142-543210 - Printed on acid-free paper

Preface and Acknowledgements

Long is the road from conception to completion. *Molière*

Geothermics, a small, but interesting and important geophysics discipline, often suffers from the fact that its potential and limits are insufficiently known by the geoscientific community at large. The work presented here, carried out in 1994/95, is intended to be a contribution toward correcting this situation.

My present work in the Geothermics and Groundwater Dynamics Section of the Joint Geoscientific Research Institute (GGA) in Hannover gave me the opportunity to undertake the research projects reported in this book. I would like to express my appreciation to my past and present division Heads and Directors, Dr. Rüdiger Schulz, Professor Dr. Ralph Hänel, and Dr, Burkhard Buttkus for their support for these projects and their interest in the problems dealt with.

Most of the examples used in this study were taken from projects in which I have enjoyed working with many colleagues too numerous to enumerate individually. However, I wish to especially thank Dr. Ernst Huenges, Dr. Ilmo Kukkonen and Prof. Jean-Claude Mareschal, who shared their enthusiasm and extensive knowledge with me.

The first version of this book, written in the years 1993 – 1995, had been accepted as DSc thesis in geophysics ('Habilitation') by the Faculty of Mathematics and Natural Sciences of the University Bonn, Germany in November 1995. The present version, translated into English by Clark Newcomb (BGR, Hannover), has been checked and updated where new data have become available.

I wish to express my appreciation to Prof. Horst J. Neugebauer for working together with me over the past years – at first informally, then in joint projects and publications, and finally in this DSc project and in teaching assignments at the university in Bonn. In numerous discussions, I have profited a great deal from his knowledge and advice.

I thank my parents whose moral and financial support was always there when it was needed throughout my entire education. And last, but not least, I especially thank my wife Karin, who not only tolerated my absence on many weekends and vacation days while this manuscript was being worked on, but also looked after me during this time and who was the first (and most critical?) reader of the manuscript. I dedicate this book to her and my children Johanna and Simon.

Christoph Clauser — Hannover, April 1999

To know that we know what we know, and that we do not know what we do not know, that is true knowledge. *Henry David Thoreau*

Contents

List of abbreviations

Symbol or Abbreviation	Definition	Unit
A	matrix of linear equation system	
α, β	compressibility of rock and fluid	Pa^{-1}
C	concentration	%
c, c_P	specific heat capacity at constant pressure	$W\ kg^{-1}\ K^{-1}$
c_f	fluid specific heat capacity	$W\ kg^{-1}\ K^{-1}$
ε	attenuation factor in the SVD	-
FD	Finite Differences	
FE	Finite Elements	
Φ	porosity	%
g	gravitational acceleration	$m\ s^{-2}$
H	heat production rate	$\mu W\ m^{-3}$
h	hydraulic head	m
h_0	hydraulic reference head	m
K	hydraulic permeability tensor	m^2
κ	thermal diffusivity	$m^2\ s^{-1}$
L	layer thickness	m
LOM	Level of Organic Metamorphism	
$\boldsymbol{\Lambda}$	diagonal matrix of the singular values	
λ	thermal conductivity tensor	$W\ m^{-1}\ K^{-1}$
m_B	basal mass flow density	$g\ a^{-1}\ m^{-2}$
μ	dynamic viscosity	Pa s
Nu	Nusselt number	
∇	Nabla operator	m^{-1}

P	pressure	Pa
Pe	Péclet number	-
$\mathbf{q}$	heat flow density vector	mW m^{-2}
q_B	basal heat flow density	mW m^{-2}
q_0	surface heat flow density	mW m^{-2}
ρ	density	kg m^{-3}
ρ_0	fluid density at reference temperature T_0	kg m^{-3}
ρ_f	fluid density	kg m^{-3}
ρ_r	relative density	-
$\mathbf{R}$	model resolution Matrix in the SVD	
R_o	average vitrinite reflectivity in oil	%
S_s	specific storage coefficient	m^{-1}
SV	singular Value	-
SVD	Singular Value Decomposition	
T	temperature	° C; K
T_{max}	maximum temperature from "Rock-Eval" pyrolysis	° C
T_0	mean temperature at the Earth's surface	° C
t	time	s
TAI	thermal Alteration Index	-
TTI	temperature Time Index	-
$\mathbf{U}$	orthonormal matrix in data space	
$\mathbf{V}$	orthonormal matrix in parameter space	
$\mathbf{v}$	Darcy velocity (specific discharge) vector	m s^{-1}
W	specific strength of hydraulic sources or sinks	s^{-1}
z	depth	m

Abstract

Many geodynamic processes are driven by the thermal energy of the Earth. They leave behind many different kinds of traces, which can be recognized and interpreted with appropriate methods. This book deals with the signatures left by these processes in the temperature field of the Earth's crust. Before concrete examples of these signatures are discussed, however, indirect methods for interpreting the present temperature field and methods for reconstructing the thermal fields of the geological past will be outlined in chapter 1.

Spatial and temporal scales that are relevant for heat transport processes are discussed in chapter 2. Of importance for local and regional interpretation is the question of how well data collected in boreholes characterize a heterogeneous subsurface. This problem is discussed both theoretically and on the basis of data from the German Continental Deep Drilling Program (KTB). Aspects of time scales related to the reconstruction of paleo-temperatures at the Earth's surface are discussed. The climate of the last 200 years and the postglacial warming in Central Europe is reconstructed on the basis of two sets of temperature data from shallow and deep boreholes.

The problem of identifying diffusion and advection components in the transport of heat in the crystalline basement is discussed in chapter 3 on the basis of data from the KTB and Kola deep borcholcs. Empirical data is interpreted, and 2-D and 3-D numerical simulations are presented. In both cases, conduction is the predominant heat transport process. This, however, does not exclude contributions from heat advection entirely.

In chapter 4, the possibilities for recognizing processes at considerable depth using data obtained at or near the surface are explored. For this purpose, an intrusion in the lower crust is modeled and the resulting thermal and flow regime is analyzed. Although thermal signals of processes at depth can be detected at the Earth's surface, ambiguities with respect to their causes can be solved only in combination with the analysis of ancillary data.

Basic equations for fluid flow and heat transport in porous media are given in the Appendix.

1 Introduction: Indicators of Past and Present Thermal Conditions

The thermal energy of the Earth is the driving force for volcanism and mantle convection, and thus for most of the associated geodynamic processes, like earthquakes, diapirism, uplift and subsidence (and the concurrent erosion and sedimentation), overthrusting, and subduction, as well as formation of grabens and mid-oceanic ridges. These processes leave thermal traces in the Earth's crust that can be studied and interpreted with the appropriate methods.

Certain minerals allow conclusions about the pressure and temperature (P-T) conditions at the time of their formation. The metamorphic grade of a metamorphic rock is indicated by the index minerals contained in it. Diagenesis and weak metamorphism are, in contrast, often indicated in sediments (especially clays) by temperature- and pressure-dependent mineral reactions. An example is the dehydration of smectite to illite. Study of xenoliths (fragments of country rock included in an igneous rock), on the other hand, yields a picture, even if somewhat blurred, of the thermal conditions in both the lower crust and upper mantle. Study of various types of rocks using the combined methods of mineralogy and metamorphic and experimental petrology thus reveals important boundary conditions for characterization of the P-T conditions for different processes, as well as at different depths (Green 1970; O'Reilly and Griffin 1990; Pearson and O'Reilly 1991; Pearson et al. 1991; Pytte and Reynolds 1989; Rudnick 1992; Thompson 1992; Tuttle and Bowen 1958; Wyllie 1988, 1992).

Metamorphic processes can lead to the release of fluids bound in the lattice of crystals. Fluids are also frequently present as fluid inclusions in minerals and in the pores of metamorphosed rock. Thus, conclusions can be drawn from the homogenization temperatures of fluid inclusions about the P-T conditions prevailing at the time of their enclosure. They are, therefore, used to reconstruct processes that take place during the formation of rocks and basins (Burruss 1989; Hollister and Crawford 1981; Roedder 1984).

Methods of isotope geology, such as fission-track analysis and other methods of radiometric age determination, permit conclusions to be drawn about the time at which an isotope or mineral system became a closed system with respect to diffusion or annealing of fission tracks. This closure age corresponds in each system to a specific closure temperature that is known from laboratory experiments. Therefore, the temperature-time (T-t) path since the time of the formation of a rock can be determined by analysis of different isotope or mineral systems of a rock. This path documents the times at which the rock was cooled below or heated above the closure temperatures of the various systems. Under certain conditions, the conditions of formation and metamorphism of the rock can be derived from this information. This, in turn, permits conclusions to be drawn about the times of uplift and/or subsidence during different orogenic cycles (England and Thompson 1984, 1986; Faure 1986; Geyh and Schleicher 1990; Haack 1977; Hawkesworth and van Calsteren 1992; Naeser 1979; Naeser et al. 1989; Thompson and England 1984; Wagner and Van den haute 1992).

Isotopic fractionation of the stable isotopes of hydrogen (D/H), oxygen ($^{18}O/^{16}O$), carbon ($^{13}C/^{12}C$) and sulfur ($^{34}S/^{32}S$), for example, is a function of temperature. If the reaction being studied was in equilibrium, conclusions can be drawn about the temperature at the time of formation on the basis of the isotope ratios in the fluids, minerals or rocks. One of the best known applications of this method is the $\delta^{18}O$ paleo-temperature scale for the $CaCO_3$-H_2O system in marine fossils containing calcium. This scale can be used to reconstruct the water temperature of paleo-oceans, at least theoretically, to ± 0.5 K. Twelve glacial periods have been identified with certainty for the last 780,000 years, the Brunhes geomagnetic epoch, using this data in combination with U/Th data and correlation with paleo-magnetic data. A further 18 – 20 have been identified for the preceeding period of up to 2 Ma before present. For the Brunhes epoch, a resolution of ± 1500 – 5000 years has been obtained in the timing of the events by correlation with different periods of the Milankovitch orbital cycle (Berger et al. 1984; Buntebarth 1984; Epstein et al. 1953; Faure 1986; Geyh and Schleicher 1990; Imbrie et al. 1984; Martinson et al. 1987; O'Neil 1979).

Several organic parameters and indices are used to describe the relationship between the T-t path followed by a sedimentary rock after deposition and the maturity of the organic matter in it. The best known are: (a) the temperature-time index (TTI); (b) the "level of organic metamorphism" (LOM); (c) the mean vitrinite reflectance (oil immersion; R_o) as a measure of the degree of coalification; (d) the EASY%R_o index for the T-t evolution of the vitrinite reflectance, based on a set of four independent, parallel reactions for the loss of water, CO_2, methane and higher hydrocarbons; (e) the thermal alteration index (TAI); and (f) the maximum temperature T_{max} obtained by pyrolysis using a "Rock-Eval" apparatus (Espitalié et al. 1977). There are also empirical relationships between R_o and LOM, R_o and TTI, and R_o and T_{max} (Hood et al. 1975; Sweeney and Burnham 1990; Teichmüller and Durand 1983; Waples 1980; Wood 1988). When certain assumptions are fulfilled, these parameters can be used to reconstruct T-t paths and paleo-temperature gradients (Barker 1989; Buntebarth and Stegena 1986; Creaney 1989; Curiale et al. 1989; Stegena 1988; Tissot et al. 1987; Tissot and Welte 1984). The quality of such reconstructions is improved by the combined application of these organic parameters with inorganicones, for instance, the results of fission-track analysis and studies of fluid inclusions (Leischner 1994; Leischner et al. 1993).

Whereas the above methods provide information about P-T conditions at different times in the geological past, “chemical geothermometers” provide information about the present temperature field. A distinction is made between strictly theoretical thermodynamic methods and empirical ones.

“Solubility geothermometers” belong to the first group, the theoretical ones: The concentrations of the individual components of a solution are a function of temperature and pressure. Therefore, under certain conditions if the system is in chemical equilibrium, the temperature can be estimated from the concentrations of the constituents of minerals that have been dissolved. Silica minerals (e.g., quartz, chalcedony and cristobalite) are commonly used for this purpose. Chemical geothermometers based on the solubility of other minerals are not frequently used, because they require that the activity coefficients of dissolved components be calculated as a function of temperature and the formation of chemical complexes be taken into consideration.

The second group of chemical geothermometers is based on the empirical relationship between the fluid temperature and the ratios of certain cation concentrations. Frequently used cation combinations are Na-K, Na-K-Ca (possibly with a Mg correction for temperatures below 200 °C), Mg-Li, and Na-Li. Chemical geothermometers cover a temperature range from 0 to 350 °C, but the ranges of application of the various geothermometers differ considerably. Above 70 °C, the quartz, Mg-Li, Mg-corrected Na-K-Ca, and Na-Li thermometers have a precision of ± 5 K. Between 30 and 70 °C, the chalcedony, Mg-Li, Na-Li and Mg-corrected Na-K-Ca thermometers yield the best results (Buntebarth 1984; Fournier 1981; Fournier and Truesdell 1973; Kharaka and Mariner 1989; Truesdell 1976).

In contrast to the indirect methods discussed above, study of the terrestrial temperature field and the heat flow density (i.e., the energy loss of the Earth per unit time and area) provides direct information about the thermal signatures of geodynamic processes. The rocks of the Earth's crust have a good thermal “memory”, owing to their comparatively poor thermal conductivity and diffusivity. This frequently permits processes in the geological past to be recognized and studied on the basis of signatures still present in the Earth's crust. Therefore, processes occurring at quite different scales can be identified and localized on the basis of the lateral variation of heat flow density (Chapman and Rybach 1985). On a global scale, these variations also provide information about the time and duration of these processes in different tectonic provinces. Moreover, the thermal energy of the Earth represents a virtually inexhaustible potential, which could be of considerable economic and ecological interest, assuming appropriate local transport conditions and possibilities for extraction and utilization.

There is a basic problem that must be solved before thermal data can be interpreted: Are the data from borehole logs representative on a local or regional scale? This is especially a problem in the case of data from the crystalline basement. The German Continental Deep Drilling Project (KTB) produced a considerable amount of new data applicable to this problem. Especially informative are the vertical temperature gradient from the borehole logs and a comparison of the densities determined from the cores with those derived from gravity measurements in the borehole. The depth of penetration of gravity and temperature gradient measurements permits a view of the rock beyond the borehole wall (Clauser et al. 1997, Kohl and Rybach 1995). The question of whether the data from borehole logs is representative, therefore, is closely related to the well known problem of scale in petrophysics and hydrogeology: How well or up to what radius are laboratory and borehole data suited for local or regional interpretation? This question applies equally for the rock properties thermal conductivity, thermal diffusivity, and radiogenic heat production rate.

For a proper understanding and interpretation of heat flow density, all processes involved in heat transport must be identified, quantified, and included in the interpretation. It is of considerable importance whether these processes are steady-state or transient. The most important heat transport processes in the Earth's crust are conduction, advection and radiation (see the Appendix for definitions of conduction, advection and convection). Only at temperatures above about 500 °C does radiation contribute significantly to the total amount of heat transported. It can then be approximated by a "radiative thermal conductivity" which, in addition to the molecular thermal conductivity, is treated by heat conduction theory (e.g., Clauser 1988b). For this reason it is not discussed in this paper as a special topic. Radiative heat transport becomes important only at temperatures above about 1000 °C, e.g., in the Earth's mantle, where radiation and advection are the prevailing transport mechanisms.

In contrast, the main means of heat transport in the Earth's crust are conduction and advection. Not too long ago, purely conductive conditions were postulated for crustal rocks in general, especially in the crystalline basement. It has been found, however, especially in studies of heat transport in sedimentary basins, that advection often makes a sizable contribution to the total heat transported (Bethke 1985; Burrus and Audebert 1990; Clauser and Villinger 1990; Düppenbecker and Welte 1992; Garven and Freeze 1984a,b; Person and Garven 1992, 1994; Smith and Chapman 1983; Ungerer et al. 1984; Welte 1982; Welte and Yükler 1981; Willett and Chapman 1987; Yalçin and Welte 1988). In the last several years, this recognition has increasingly led to associating vertical heat flow density variations in the crystalline basement with advection of heat by convection of fluids in the crust (Clauser and Neugebauer 1991; Forster and Smith 1989, 1990). Especially the quality and density of the data from deep research boreholes in the crystalline basement permit detailed study of these phenomena (Jobmann and Clauser 1994; Kohl and Rybach 1995; Kukkonen and Clauser 1994). However, unambiguous correlation of vertical variations in heat flow density and the temperature gradient with individual processes is not straightforward. On the one hand, in a plot of the vertical components of the heat flow density and the temperature gradient versus depth, deviations from a straight line similar to those caused by advection are produced by both (i) refraction of the vertical flow of heat resulting from laterally heterogeneous and anisotropic thermal conductivity and (ii) transient downward propagation by thermal diffusion of climatic temperature variations at the Earth's surface. On the other hand, interpretation of thermal data together with other physical or chemical parameters, e.g., the density of the rock or isotopic ratios, provides a means of reducing this ambiguity. Thus, it may be expected that, for instance, the thermal "memory" of the Earth's upper crust can be utilized at appropriate places to reconstruct paleo-temperature variations at the Earth's surface as a function of time directly from the thermal traces left in the crust by the downward diffusion of this signal. Prerequisite for this, however, is that the corresponding signals in the temperature field can be correctly identified and correlated with the associated processes.

The problems described above can be divided into two groups: (1) problems related to spatial and temporal scales and (2) the problems of differentiating between different heat transport processes. Examples from these two groups are discussed in the following three chapters. Problems related to spatial and temporal scales are discussed in chapter 2: On the one hand the KTB borehole is used as an example of steady-state heat conduction in a heterogeneous subsurface, and the question is delved into as to how representative borehole data is with respect to the local and regional scales. On the other hand the reconstruction of paleo-temperatures at the Earth's surface from temperature logs recorded in shallow and deep boreholes of the KTB project is discussed. Differentiation between steady-state heat conduction, advection and transient diffusion is discussed in chapter 3 on the basis of two case studies related to the two deepest boreholes in crystalline basement: the Kola and the KTB super-deep boreholes. The identification of processes at depth is discussed in chapter 4 on the basis of a numerical experiment on the relationship between crustal dynamics and transient thermal advection in continental rift zones.

2 Spatial and Temporal Scales for Heat Transport Processes

The spatial and temporal scales that describe heat transport by the two most important transport processes in the Earth's crust, diffusion and advection, can differ considerably from each other. The reason for this difference is to be found in the different local possibilities for these two processes to occur, as well as differences in their velocities.

Diffusion is a ubiquitous, relatively slow process in the Earth's crust. A mean thermal diffusion constant κ of 10^{-6} m^2 s^{-1} = 1 mm^2 s^{-1} $\approx$ 31.56 m^2 a^{-1} is commonly used for rocks. κ is a function of thermal conductivity λ, density ρ, and specific heat capacity at constant pressure c_P: $\kappa = \lambda/(\rho c_P)$. As a result of this relationship, there is normally less variation in thermal diffusivity than in thermal conductivity (Beck 1988).

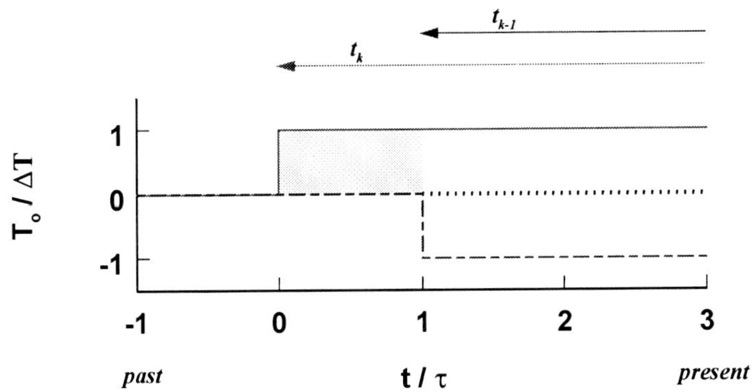

Fig. 2.1. Rectangular variation of temperature T_0 at the surface of a half-space resulting from two opposing temperature steps of amplitude $\pm\Delta T$ at times 1 and 2 normalized with respect to the signal length τ.

The following example is discussed to illustrate both the spatial and temporal scales for diffusion in the Earth's crust: Figure 2.1 shows the mean temperature T_o at the surface of a half-space plotted versus time. T_o is composed of two successive, equal, but opposite temperature steps of amplitude ΔT occurring at an interval of τ time units apart. The propagation of this thermal signal with time downwards from the surface into a homogeneous, isotropic half-space with a thermal diffusivity κ is described by the following equation (Carslaw and Jaeger 1959, p. 63):

$$\frac{T(z,t)}{\Delta T} = \operatorname{erfc}\left\{\frac{z}{2\sqrt{\kappa\, t_k}}\right\} - \operatorname{erfc}\left\{\frac{z}{2\sqrt{\kappa\, t_{k-1}}}\right\} , \tag{1}$$

where erfc is the complementary error function.

The times t_k and t_{k-1} represent the beginning and end of the rectangular signal with a duration $\tau = t_k - t_{k-1}$. The spatial and temporal temperature distribution obtained from equation (1) is shown in Plate 1, normalized with respect to the signal amplitude ΔT and on the time axis with respect to the signal length τ. The scale of the depth axes depends on the signal duration τ, where the depth is given by multiplication of the scale by $\tau^{1/2}$. This is particularly easy for integer powers of 10 (τ = 1, 100, 10,000, ... years for the left axis and τ = 10, 1000, 100,000 years for the right axis). This nomogram, modified after Lachenbruch and Marshall (1986), shows the maximum depth to which signals of different durations will penetrate at a given percentage level. For instance, signals of events with a 10-year and 100-year duration will be attenuated to 15 % of their original amplitudes at depths of about 45 m and 135 m, respectively. The dashed line in Plate 1 shows the depth of the temperature signal's maximum with respect to time. This curve is obtained by differentiation of equation (1) with respect to depth and setting the derivative equal to zero:

$$z(T_{max}, t) = \sqrt{\frac{2\kappa(t_k \cdot t_{k-1})\ln(t_k/t_{k-1})}{(t_k - t_{k-1})}} \tag{2}$$

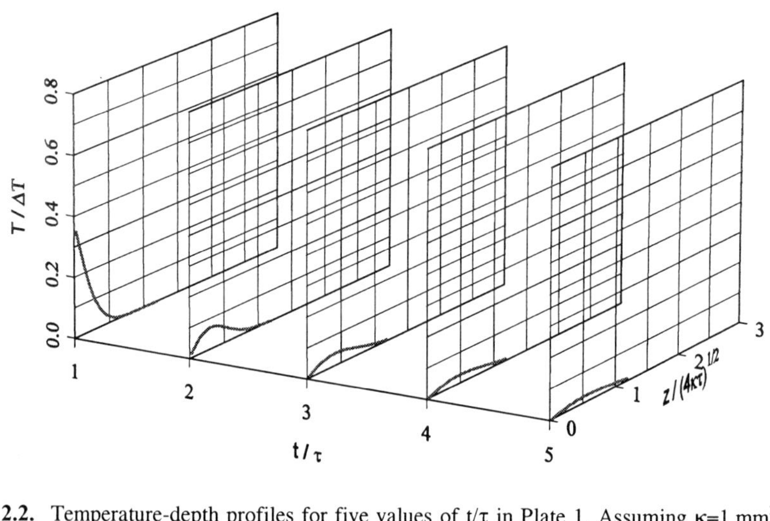

Fig. 2.2. Temperature-depth profiles for five values of t/τ in Plate 1. Assuming κ=1 $mm^2\ s^{-1} \approx$ 31.56 $m^2\ a^{-1}$, the normalizing factor for the dimensionless depth axis equals about 11.23 $\tau^{1/2}$.

The usual temperature logs are obtained by taking vertical sections through this plot for different times (Fig. 2.2). At the end of a rectangular pulse ($t/\tau = 1$) of $\tau = 1$ year (or 100 or 10,000 years), the signal below about 15 m (or 150 or 1500 m) is already less than 5 % of its initial value (Plate 1). For signal durations τ of 10 years (or 1000 or 100,000 years), the corresponding depths are 50 m (or 500 or 5000 m). Two signal lengths after the end of the rectangular pulse (i.e., $t/\tau = 3$), the signal has been attenuated to less than 10 % of its initial amplitude throughout the entire half-space. The maximum of 10 % is reached at a depth of about 12 m (or 120 or 1200 m) for a signal duration of $\tau = 1$ year (or 100 or 10 000 years) and at about 39 m (or 390 or 3900 m) for $\tau = 10$ years (or 1000 or 100,000 years). These spatial and temporal scales are characteristic of the propagation of thermal diffusion processes in the Earth's crust.

Advective heat transport always requires a flow system that is capable of passing on a thermal signal. Therefore, the specific discharge associated with the flow system, the Darcy velocity, determines the thermal efficiency of advection, and not a petrophysical property such as thermal diffusivity. The Darcy velocity is defined as follows:

$$\mathbf{v} = \frac{\underline{\mathbf{k}}}{\mu}(\nabla P + \rho g \nabla z) \ , \qquad (3)$$

where $\underline{\mathbf{k}}$ is the tensor of the rock permeability, μ, ρ and P are the dynamic viscosity, density and pressure of the fluid, g is the gravitational acceleration, and z is the depth. Accordingly, the Darcy velocity is governed by the hydraulic resistance, which is determined by the permeability of the rock, the viscosity of the fluid, and the driving forces derived from the pressure gradient and thermal buoyancy. Hence, in contrast to diffusion, several parameters describe advective transport with respect to space and time. Therefore, a parameter is defined that expresses the effects of these variables and at the same time normalizes advective transport with respect to diffusive transport. Such a parameter is the thermal Péclet number, Pe, a dimensionless normalizing factor resulting from solving the advection-diffusion equation. In the simplest one-dimensional, steady-state case, it is defined as follows (see, for example, Clauser and Villinger 1990):

$$Pe = \frac{q_A}{q_D} = \frac{(\rho c)_f \, v \Delta T}{\lambda \left(\dfrac{\Delta T}{L} \right)} = \frac{(\rho c)_f \, v L}{\lambda} \qquad (4)$$

where λ is the bulk thermal conductivity of the fluid-saturated rock, $(\rho c)_f$ is the thermal capacity of the fluid, L is the typical transport distance, and ΔT is the temperature difference within this distance; v is considered to be constant, and the negative v-axis is downwards.

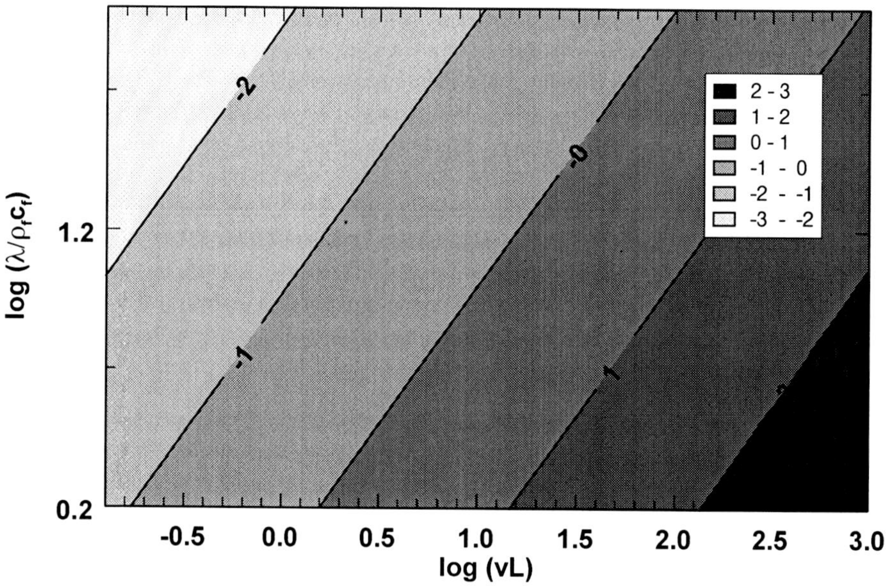

Fig. 2.3. The Péclet number, defined by eq. (4), as a function of $\lambda/(\rho c)_f$ and vL

Both vL and $\lambda/(\rho c)_f$ have the dimensions [Length^2 / Time] and can therefore be considered formally as "diffusion constants" for heat advection and conduction, respectively. Pe is the ratio of the two parameters and thus a measure of the relationship between advective and diffusive heat transport q_A and q_D. The Péclet number is illustrated in Figure 2.3 as a function of the two parameters vL and $\lambda/(\rho c)_f$. The unit of time used for v and λ when plotting $\lambda/(\rho c)_f$ vs. vL in Fig. 2.3 is one year (i.e. [v]= m a^{-1}; [λ] = J a^{-1} m^{-1} K^{-1}, with 1 a = 31,557,600 s). Taking the commonly used value of 2.3 MJ m^{-3} K^{-1} for $(\rho c)_f$ (e.g., Beck 1988) to calculate Pe, the normal range of values for the thermal conductivity of rocks ($0.5 \leq \lambda \leq 7$ W m^{-1} K^{-1}) corresponds in Figure 2.3 to the interval from about 0.8 to 2.0 on the y-axis. An abscissa value of 1, for example, corresponds to v = 1 mm a^{-1} and L = 1 km. These are typical values for both the crystalline basement and sedimentary basins. Between the points (1,0.8) and (1,2.0), the Péclet number assumes values between 0.1 and 1. The transition between predominantly diffusive and predominantly advective heat transport occurs in this range. The corresponding range of Péclet numbers for other combinations of Darcy velocity v and transport distance L can be taken from Figure 2.3.

Advective heat transport is often also quantified using the Nusselt number Nu, which is likewise dimensionless. This parameter is the ratio of total heat transported to the conductive component. The relationship between the Péclet and Nusselt numbers is given by equation (5):

$$\mathrm{Nu} = \frac{q_A + q_D}{q_D} = \frac{q_A}{q_D} + 1 = \mathrm{Pe} + 1 \ . \tag{5}$$

2.1 How Representative are Borehole Data for a Heterogeneous Rock Mass With Respect to Local and Regional Scales?

In contrast to seismic, electric, magnetic and gravity studies, geothermal studies are based primarily on data obtained from boreholes: temperature logs and thermal properties derived from cores. Temperature logging, like gravity logging, is a "passive" method, i.e., one that does not itself transmit a signal into the rock. Thus, in contrast to "active" methods, like in sonic, electric, or nuclear logging, there is no "depth of penetration" as such. But there is a radius around a borehole in which differences in the thermal properties of the rock have a measurable effect on temperature and temperature gradient in the borehole. Obviously, structures near the borehole have a larger influence than those further away, but depending on the differences in their thermal properties, the latter can also have recognizable effects on the temperature and temperature gradients: In a certain sense, the thermal parameters "look" beyond the borehole.

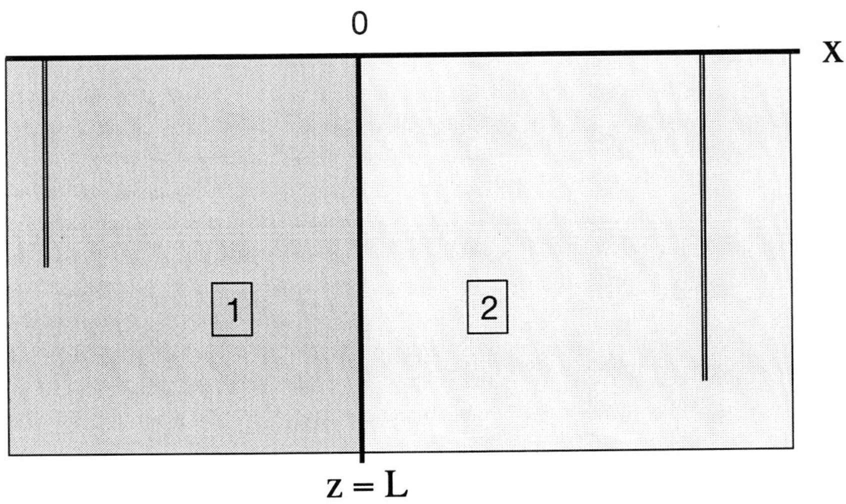

Fig. 2.4. Vertical contact between two media of thickness L with different thermal properties. Two boreholes are located on both sides at some distance from the contact.

This leads, on the one hand, to the problem of identification of heterogeneities around a borehole on the basis of temperature logs. On the other hand, there is the question of the minimum rock volume in which average thermal properties of the rock are sufficiently representative to describe the temperature field and temperature gradient - analogous to the "representative elementary volume" (REV; Bear 1972) known from hydrogeology. This is particularly important in highly heterogeneous media, in which the thermophysical properties of the rock in the drill core can be completely different from those in the surrounding rock, e.g., in the extreme case of very steeply dipping strata. This has significant consequences for the determination of heat flow density. The theoretical aspects of both problems are discussed in this section using an example from the KTB main borehole.

The influence of differences in thermal conductivity on a temperature log measured some distance from the vertical contact between two media is illustrated here for the steady-state case on the basis of the simple arrangement shown in Figure 2.4. Thermal conductivities of 1 and 3 W m^{-1} K^{-1} are assumed for the two media, undisturbed vertical temperature gradients of 60 and 20 mK m^{-1}, a constant thermal capacity ρc of 2.3 MJ m^{-3} K^{-1}, and a constant heat flow density of 60 mW m^{-2}. The maximum temperature difference ΔT_{max} that can exist between the two boreholes at a depth z is defined by the undisturbed regional heat flow density q and the harmonic mean of the two thermal conductivities (e.g., Poley and van Steveninck 1970):

$$\Delta T_{max}(z) = qz \frac{\lambda_1 - \lambda_2}{\lambda_1 \lambda_2} \ . \qquad (6)$$

The temperature at the contact between the two media (x = 0) is given by the following equation (Hänel and Mongelli 1988):

$$T_{contact}(z) = T_o + z \frac{\sqrt{\lambda_1 \rho_1 c_1}\left(\frac{\partial T}{\partial z}\right)_1 + \sqrt{\lambda_2 \rho_2 c_2}\left(\frac{\partial T}{\partial z}\right)_2}{\sqrt{\lambda_1 \rho_1 c_1} + \sqrt{\lambda_2 \rho_2 c_2}} \ , \qquad (7)$$

where T_o is the mean temperature at the Earth's surface. T_{min}, T_{max}, and $(\partial T/\partial z)_{1,2} = q/\lambda_{1,2}$ are the undisturbed temperatures and the vertical temperature gradient in the two media at a sufficient distance from the contact. Figure 2.5 shows the results for equation (6) and (7) using the parameter values discussed above and T_o = 0 °C. Carslaw and Jaeger (1959, p. 428) give the following temperature distributions for the right and left sides of the contact ($|x| > 0$):

$$T_1(x<0,z) = T_o + z\frac{q}{\lambda_1} + \frac{2qL}{\pi}\frac{(\lambda_2-\lambda_1)}{\lambda_1(\lambda_1+\lambda_2)}\sum_{n=1}^{\infty}\frac{(-1)^n}{n}\sin\frac{n\pi z}{L}\exp\frac{n\pi x}{L}$$

(8)

$$T_2(x>0,z) = T_o + z\frac{q}{\lambda_2} + \frac{2qL}{\pi}\frac{(\lambda_2-\lambda_1)}{\lambda_2(\lambda_1+\lambda_2)}\sum_{n=1}^{\infty}\frac{(-1)^n}{n}\sin\frac{n\pi z}{L}\exp\frac{-n\pi x}{L} .$$

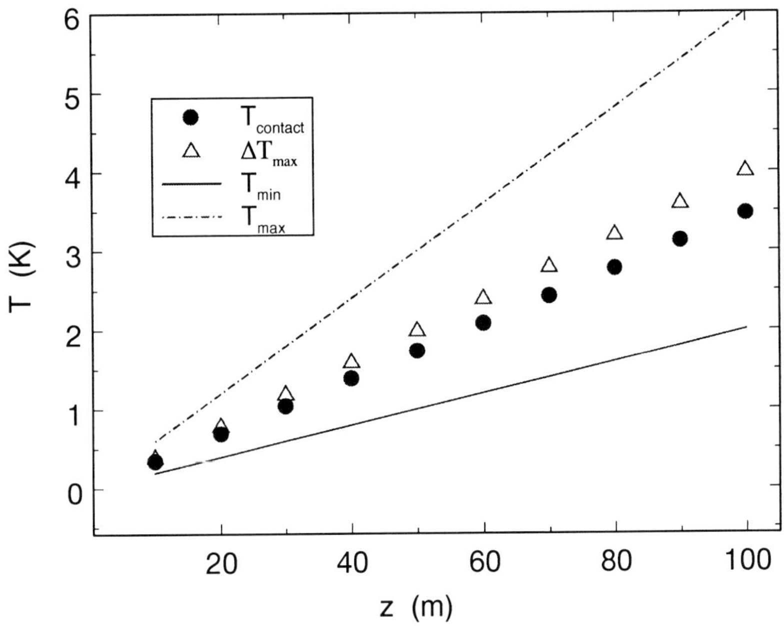

Fig. 2.5. Temperature-depth profile around a vertical contact between two media with different thermal conductivities: ΔT_{max} and $T_{contact}$ are calculated from equations (6) and (7); T_{max} and T_{min} are calculated from from $T_{contact} \pm \Delta T_{max}$, respectively.

In Figure 2.6 temperature is shown as a function of depth and distance from the contact ($x = 0$). The configuration is the one shown in Figure 2.4 with $L = 2000$ m and the parameter values given above. Equation (7) was used to calculate the temperatures at the contact; equation (8) was used for the remaining areas. The temperatures at the lower edge of regions 1 and 2 ($z = L$) are 40 °C and 120 °C, respectively, in accordance with the constant heat flow density and the thermal conductivities of the two regions. T_{min}, T_{max} and the two undisturbed vertical temperature gradients of 20 and 60 K km^{-1} can be seen at the left and right sides of the figure.

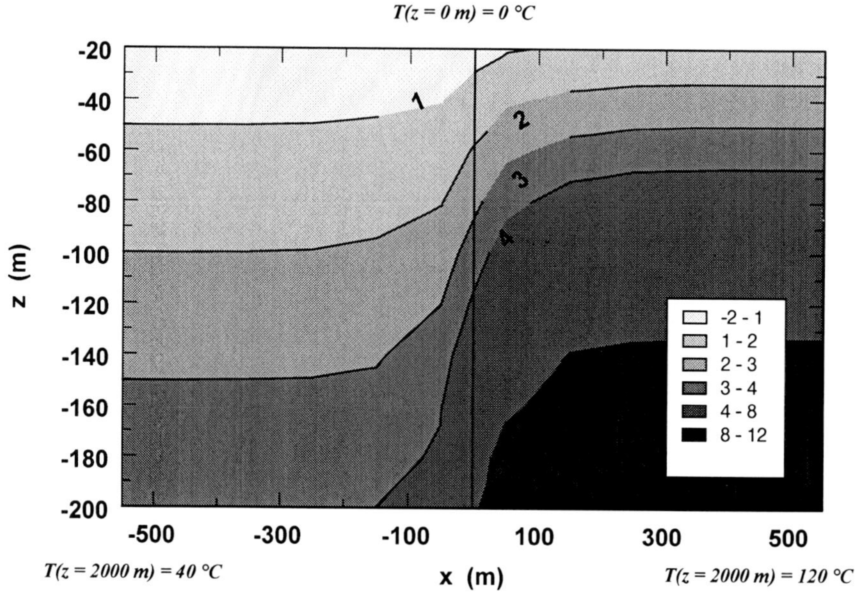

Fig. 2.6. Temperature distribution to the left and right from a vertical contact between two media of different thermal conductivities, calculated according to equation (8).

Geometries more complicated than the one discussed above can be handled with numerical methods. Chen and Beck (1991), for example, used the method of boundary elements to study the influence of an elliptical, two-dimensional rock mass (axes: 5 and 15 m) near a borehole on temperature and temperature gradient in the borehole itself. The thermal conductivity of this elliptical cylindrical mass is 10 % of that of the surrounding material; its orientation varies from 0° to 90° and its distance from the borehole varies from 1 to 5.4 m. In contrast to the previous examples, the rock mass causing the thermal disturbance is relatively near the borehole. Corresponding to the size of the rock mass, the large thermal conductivity difference between the cylindrical rock mass and the surrounding rock, and the proximity to the borehole, the temperature and temperature gradient signals are relatively large: ± 3.5 K and -0.2 – 1.2 K m^{-1}. An unrecognized rock mass with a high thermal conductivity difference with respect to the surroundings can thus lead to considerable variation in the temperature and vertical temperature gradient. A difference on this order of magnitude is not impossible for rocks, but it is the exception.

Although these results are, strictly speaking, valid only for the selected parameter values, they illustrate, however, the influence of variations in thermal conductivity in the vicinity of a borehole on the T log. Signals associated with differences smaller than those of the examples are correspondingly smaller. This makes them more difficult to identify. If the signals are attributed to other effects or transport mechanisms, misinterpretation is inevitable. On the other hand, if they are identified, thermal methods make it possible to look beyond the immediate vicinity of a borehole.

This can be illustrated using the two KTB boreholes as an example. The results of the thermal measurements in the pilot and main boreholes are shown in Plates 2 and 3. The figures show from left to right: (1) Temperature T and reduced temperature $T_{red} = T(z) - 7.5 - 0.0275\,z$ (where 7.5 °C is the mean annual temperature at the Earth's surface and 0.0275 K m^{-1} is the mean vertical temperature gradient; lines marked by a ⊕), bottom-hole temperature (BHT) extrapolations (red full circles) as well as reduced BHT extrapolations calculated as described above (circled crosses with uncertainty range indicated by a horizontal red bar); (2) The vertical temperature gradient $\Delta T/\Delta z$; (3) The thermal conductivity λ; (4) The vertical heat flow density $q_z = \lambda\ \Delta T/\Delta z$; (5) The contribution of radiogenic heat production rate H to q_z; (6) A highly simplified lithological column for each borehole according to the three main types of lithological units of the drilled sections: gneiss (red), metabasite (green) and an alternating sequence of these two (orange). In Plate 3, the rock density ρ in the main borehole, determined from borehole logging and laboratory measurements on core, is additionally shown in column 5.

Two temperature logs (blue and red lines) from the pilot borehole are shown in the first column of Plate 2, together with seven bottom-hole temperature (BHT) extrapolations (red full circles) and reduced BHT extrapolations (circled crosses with red uncertainty range). Yellow shading indicates the areas above and below about 2.3 km that have been heated or cooled by the drilling mud. The log in blue was recorded on 23 November 1989, about eight months after completion of the drilling to a final depth of 4 km. At this time, the thermal perturbation caused by the drilling had decayed to a large degree, even if not completely. The log in red was recorded on 17 September 1997 about 8 ½ years after the end of drilling operations. At this time, the rock around the borehole had largely reattained thermal equilibrium. The BHT extrapolations shown on the graph are partly based on extremely long observation times of up to 663 hours. The maximum difference between these values and the temperature logs is about 1 K. This is very good agreement, in view of the highly simplified assumptions on which the BHT extrapolations are based. The difference between the earlier and the later T logs, caused by the gradual decay of the thermal disturbance with time, can be clearly seen in the reduced temperature curves (⊕). After 8 ½ years the thermal perturbation has decayed to a large degree.

The vertical component of the temperature gradient (column 2) was calculated for each 5 m as a moving average over depth intervals of 50 m. The difference between the two temperature gradient logs in blue and red, calculated from the early and late temperature logs, respectively, show up clearly here, too. They are due to the fading of the thermal disturbance caused by the drilling operations. The vertical component of thermal conductivity (column 3) was determined on the basis of measurements on drill cores in the KTB field laboratory (Huenges et al. 1990;

Berckhemer et al. 1997). The points represent the values measured about 5 m apart and the line represents the mean, calculated like for the temperature gradient as a moving average over depth intervals of 50 m. The shaded areas represent mean values plus and minus one standard deviation for the depth intervals corresponding to the lithological log.

The fourth column of Plate 2 shows the vertical component of the heat flow density calculated using different methods: The blue and red curves are the product of the corresponding two mean vertical temperature gradients in column 2 and the mean vertical thermal conductivity in column 3. The areas shaded in blue represent the product of the mean in column 2 for the depth ranges of the lithological log and the upper and lower limits of the shaded areas in column 3. The yellow area in the middle shows the effect of the temperature dependence of thermal conductivity on heat flow density calculated using incremental "Bullard Plots" (Bullard 1939; see e.g. Clauser and Huenges 1993): The left boundary of the yellow area corresponds to the heat flow density calculated taking into consideration the temperature dependence according to Buntebarth (1991); The right boundary corresponds to the value calculated not taking this dependence into account.

The fifth column shows the cumulative contribution of the radiogenic heat production rate to the heat flow density, and the sixth column shows the three main lithological units drilled in the KTB pilot borehole: gneiss (red), metabasite (green) and alternating gneiss and metabasite (orange).

Plate 3 illustrates the corresponding results for the main KTB borehole. The first column shows the temperature T (cyan, green, magenta, and red lines) and the reduced temperature $T_{red.}$ (calculated as for Plate 2; line marked by ⌖ in the figures) determined from eight BHT extrapolations (red full circles) and reduced BHT extrapolations (circled crosses with uncertainty range). The logs were recorded in the main hole in January 1997 (red), October 1994 (cyan), August 1993 (magenta), and February 1993 (green). Also shown in blue are the corresponding curves of the temperature log recorded in September 1997 in the pilot borehole (see Plate 2). Owing to the unfavorable conditions (borehole diameter three times that of the pilot borehole, a maximum of only 188 hours shut-in time), there is a larger uncertainty for the BHT extrapolations in the main borehole than in the pilot borehole. Comparison of the BHT extrapolations and the temperature logs made immediately on completion of drilling shows the depth intervals warmed or cooled by the drilling mud (yellow shading). While the T logs recorded immediately after shut-in indicate a maximum cooling of about 60 K due to the drilling operations, this disturbance is less than 5 K after two years of shut-in. Full equilibration will require considerably more time.

The next four columns of Plate 3 show the mean values and confidence intervals for the vertical component of the temperature gradient (column 2), the thermal conductivity (column 3), the vertical component of the heat flow density (column 4), the cumulative contribution of the radiogenic heat production rate to the heat flow density as well as the rock density ρ in the main borehole (column 5).

Rock density was determined both from borehole gravity measurements (red line) and on drill cuttings in the laboratory (green dots). The sixth column shows the three main lithological units drilled in the KTB main borehole: gneiss (red), metabasite (green) and alternating gneiss and metabasite (orange).

Yellow shading in columns 3 and 4 indicates the decrease of thermal conductivity with temperature and the associated reduction in heat flow density. In contrast to the pilot borehole, mainly drill cuttings and few drill cores are available from the main borehole. The value for thermal conductivity in this case is a mean for all principal directions in this anisotropic medium, because thermal conductivity values obtained from unoriented drill cuttings cannot be used to determine tensor components.

Below about 1.6 km, the temperature gradient is constant and equal to about 27.5 K km^{-1} within its limits of error and shows no significant decrease with depth. Below a depth of about 3 km, the thermal conductivity drops to low values that correspond to the metabasite drilled through, whereas above 3 km the higher thermal conductivity of gneiss appears to predominate. The heat flow density reflects, as product of temperature gradient and thermal conductivity, the relatively large uncertainties. For this reason it also has a constant mean value (85 mW m^{-2}) within the limits of error over the total depth interval. Interestingly, there is a large drop to lower values at a depth of 3 km. This is caused by the change in thermal conductivity at this depth which is not compensated by a corresponding, opposite change in the temperature gradient, as would be expected for purely vertical, conductive heat transport. The variations in both the temperature gradient and the heat flow density correlate with the alternations between metabasite and gneiss, due to differences in their thermal conductivities. Owing to the steep dip and foliation of the rock sequence, this correlation is understood as an indication of lateral heat transport: Within the depth interval 3 - 7 km the borehole is almost entirely within a steeply dipping metabasite layer. Thermal conductivity values determined from drill cuttings are correspondingly low. These metabasite dikes, however, usually do not have a significant lateral extent and are embedded in the predominant gneiss. Therefore, if gneiss predominates in the area around the borehole, the temperature gradient reflects its higher thermal conductivity (see also Fig. 2.6). This would already sufficiently explain the drop in the heat flow density to lower values between 3 km and 7 km, considering that the difference between the mean thermal conductivities of gneiss and metabasite in the KTB pilot borehole is about 1.3 (Pribnow 1994; Clauser and Huenges 1995) (Fig. 2.7). However, this offset between 3 km and 7 km does not occur in the temperature gradient. The component of the temperature gradient normal to the boundary between two materials of different thermal conductivities is discontinuous, whereas the component of the temperature gradient parallel to this boundary is continuous. The heat flow density components show just the opposite behavior in this regard. The simultaneous presence of an offset in the heat flow density at a depth of 3 km and the lack of such an offset in the temperature gradient indicates lateral variations in thermal conductivity that are not reflected in the values determined on drill cuttings or cores (Clauser et al. 1997). Kohl and Rybach (1995) confirm this interpretation using detailed, two-dimensional numerical models based on the geological model of the area around the KTB borehole by Hirschmann (1993). Support for this interpretation, which is independent of thermal parameters, is given by the difference in the two density logs in the fifth column of Plate 3. Above a depth of about 3 km - in a section consisting predominantly of gneiss and alternating gneiss and metabasite (column 7) - the density values determined on drill cuttings (green dots) and from borehole gravity logging (red line) are in good agreement. Below that depth, they are significantly different: The density values determined on drill cuttings are higher, corresponding to the density of the metabasite being drilled through. In contrast, the interpretation of the borehole gravity log yielded lower values that are more typical of gneiss.

This example illustrates how a joint interpretation of different data can lead to conclusions about the conditions around a borehole, even when the geology is complicated. Thermal methods, combined with suitable ancillary information, make it possible, so to speak, to look beyond the immediate vicinity of a borehole.

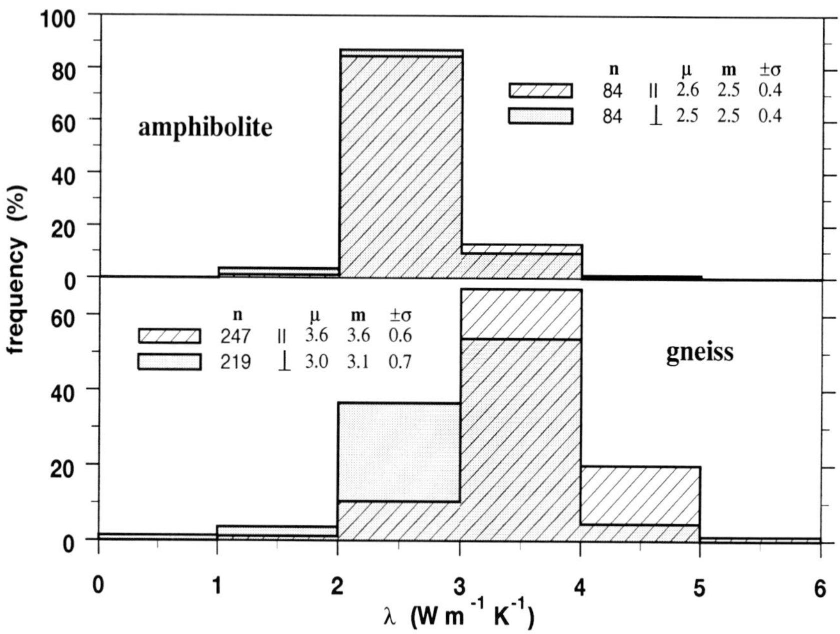

Fig. 2.7. Frequency distribution of the thermal conductivity of metabasite (amphibolite) and gneiss from measurements made with a half-space line source parallel (‖) and perpendicular (⊥) to the foliation in drill core samples from the KTB pilot borehole (Clauser and Huenges 1995); n is the number of samples, μ and m are the mean and median, and σ is the standard deviation.

2.2 Reconstruction of Paleo-Temperature at the Earth's Surface

The thermal conditions in the Earth's crust are not constant. At any particular time, the thermal regime reflects the tectonic conditions and the prevailing orogenic cycle of that time. On the other hand, the previous conditions have left traces that can be detected, e.g., in the P-T-t paths determined from isotope and mineral systems, in fluid inclusions and fission tracks, as well as in a multitude of organic parameters. This field of paleo-geothermics, which is not the subject of this book, is briefly described in chapter 1. Paleo-geothermal methods are used to determine the thermal conditions in the Earth's crust in the geological past.

Climatologists, in contrast, study the changes in temperature that have occurred in the geological past at the Earth's surface, in the atmosphere, and in the oceans. Some of the most important geoscientific methods associated with this subject are touched on briefly in chapter 1, e.g., isotope methods and analysis of Milankovitch cycles (periodic variations in the precession of the Earth's axis, in the eccentricity of the Earth's orbit around the sun, and in the time at which the Earth reaches its perihelion). Other methods involve the study of varves (annual layers in lakes) and their biostratigraphy, which permit dating on the basis of cycles in laminated clay or the micro-paleontological content of the lake sediments, as well as palynology and dendrochronology (the study of the pollen content of continental sediments and of the growth rings of trees) (Zolitschka 1990; Negendank and Zolitschka 1993; Faegri and Iversen 1989; Moore et al. 1991; Schweingruber 1988). Whereas all these methods sometimes achieve a quite remarkable temporal resolution (using palynological methods or by studying varves, for example, even seasonal changes can be identified on the basis of the pollen or microfossil content), they only permit an indirect conclusion about temperature. They are, therefore, termed paleo-climatological data or "proxy data". In contrast, the temperature signals associated with the paleo-climate are preserved themselves in the Earth's crust as a result of the low thermal diffusivity of the rocks. Study of these signals therefore provides direct access to absolute values of the paleo-climatic temperature variations at the Earth's surface. Owing to diffusion, however, sharp time boundaries cannot be expected.

2.2.1 Inversion of Borehole Temperature Data

It has been known for some time that temperature variations at the Earth's surface in the past are preserved as disturbances in the present steady-state subsurface temperature field (Lane 1923). Such transient fluctuations of the surface temperature of the Earth are attenuated exponentially with depth. They penetrate to a depth which depends on the frequency of the temperature signal: The annual wave is attenuated to 1/e ≈ 37 % of its original value already at a depth of 8 m, a signal with a period of 100 years at a depth of 80 m. At depths of 25 m and 250 m, respectively, their amplitudes are less than 5 % of their initial value (Plate 1). Temperature logs are often recorded in boreholes to determine heat flow density. But because the end of the Pleistocene glacial epoch and the climate history of the Holocene have led to large transient disturbances of the temperature gradient, corrections need to be made to eliminate these effects on the measured steady-state heat flow density (e.g., Benfield 1939; Bullard 1939; Birch 1948; Jessop 1971; Kappelmeyer and Hänel 1974; Clauser 1984). To this end, a curve for post-glacial temperature is assumed for the corrections of the temperature gradient and heat flow density. Beck (1977) pointed out that because temperature measurements are made more frequently in shallow boreholes (< 500 m) than in deep ones, the temperature variations during the Holocene have a larger influence on the calculated temperature gradient than those at the end of the Ice Age, and that the temperature history of the last 1000 years has the largest influence on the temperature gradients determined from these temperature logs. The inverse problem, determination of the ground temperature history (G.T.H.) from borehole temperature logs, was first formulated by Hotchkiss and Ingersoll (1934). Beck and Judge (1969) and Beck (1982) interpreted the disturbances in two shallow temperature profiles in southern Ontario as an indication of recent, regional warming. Čermák (1971) determined the ground temperature history near Kapuskasing in northern Ontario from borehole temperature logs. Vasseur et al. (1983) were the first to use inversion algorithms to reconstruct the ground temperature history from temperature logs. They determined the ground temperature history for the Holocene in the Limousin region of central France. The general interest in borehole temperature logs was further increased by the studies of Lachenbruch and Marshall (1986), which indicated a warming of 1 – 2 K in Alaska over the last 100 years. They postulated that this could be the first sign of an increased greenhouse effect, as predicted by global circulation models. This interpretation is at least consistent with the analysis of global temperature records, which indicate considerable warming of the arctic regions (Hansen and Lebedeff 1987).

For this reason, reconstruction of ground temperature histories from borehole temperature logs has been the objective of a number of studies in the last several years (e.g., Nielsen and Beck 1989; Mareschal and Beltrami 1992; Shen and Beck 1991; Wang 1992). Temperature logs from different parts of the world have been studied by various researchers (e.g., Lewis 1992). Work in Canada has found a warming of 1 – 2 K over the last 150 years for the eastern part of the country (Beltrami et al. 1992), preceded by a colder phase (Wang and Lewis 1992). The effect of the clearing of virgin forests on ground temperature history has been studied in Cuba (Čermák et al. 1992) and Zaire (Sebagenzi et al. 1992). In Europe, a study of two boreholes in France has revealed three main phases in the ground temperature history of the last 1000 years: the climatic optimum in the Middle Ages, the "Little Ice Age" around 1700, and the present warming (Mareschal and Vasseur 1992). A study of the Bohemian Forest area shows an apparently abrupt warming 480 years ago that correlates with the clearing of forests (Šafanda and Kubik 1992). The objective of all these studies is to supplement the data from meteorological records and paleo-climatological studies. If successful, the temperature history at the Earth's surface could be determined from temperature logs in a large number of boreholes on the continents for a period of time that extends further into the past than most instrumental records. Following a brief introduction to the basic theory, this will be illustrated on the basis of two sets of data for two different periods of time: The T log from the KTB pilot borehole contains information about the end of the Würm glacial stage about 10,000 years ago, and data from nine shallow boreholes in Bavaria and six in Bohemia permit reconstruction of the climate history of Central Europe back to about the year 1700.

In a horizontally layered, purely conductive rock, the steady-state equilibrium temperature $T_e(z)$ is described by

$$T_e(z) = T_0 + q_0 R(z) - M(z) \ . \tag{9}$$

T_0 is the equilibrium reference temperature and q_0 is the reference heat flow density at the Earth's surface, R(z) is the integrated thermal resistance down to a depth z, and M(z) is a term that takes heat production into consideration:

$$R(z) = \int_0^z \frac{dz'}{\lambda(z')} \quad \text{and} \quad M(z) = \int_0^z \frac{dz'}{\lambda(z')} \int_0^{z'} H(z'')\,dz'' , \tag{10}$$

where λ(z) is thermal conductivity and H(z) is the heat production rate. Data on λ and H are obtained from measurements on drill cores or estimated from lithology.

A number of factors can influence the shape of a temperature profile, for example, deviations from the assumed one-dimensional structure (e.g., lateral changes in thermal conductivity, heat production rate, or surface temperature, and topographic effects), non-conductive heat transport, or deviations from steady-state thermal equilibrium resulting from transient variations in surface temperature. If all other factors that influence a temperature log can be excluded, then deviations from a piecewise-linear temperature curve in a T log can be attributed to changes in the surface temperature. In this case, the subsurface temperature field is a superposition of an equilibrium state T_{eq} and a transient perturbation T(z,t). The inverse problem requires the simultaneous determination of the reference heat flow density q_0 and the transient boundary condition T(z=0,t) for the Earth's surface temperature, i.e., the ground temperature history (e.g., Vasseur et al. 1983; Shen and Beck 1983; Nielsen and Beck 1989; Shen and Beck 1991; Wang 1992). The inversion method used here is the "generalized inverse" or "singular value decomposition" (Lanczos 1961; Jackson 1972; Menke 1989). Its application to the geothermal inverse problem is described in detail by Mareschal and Beltrami (1992) and Clauser and Mareschal (1995). The solution of the 1-D diffusion equation using initial and boundary conditions for a conductive half-space provides a first approximation for the temperature disturbance (Carslaw and Jaeger 1959):

$$\frac{\partial T}{\partial t} = \kappa \frac{\partial^2 T}{\partial z^2} , \tag{11}$$

where t is time, κ is thermal diffusivity, and z is depth, with the positive axis downwards. The effect of heat sources or sinks that vary with time, such as in permafrost regions, is not accounted for in equation (11). This 1-D approximation can be used only if the spatial wavelength of the long-term fluctuations in surface temperature is large with respect to the depth of penetration of these changes (a maximum of several km). Thermal diffusivity can vary with depth, but unless there is a really large difference, these changes cause only a small disturbance in the transient signal. For this reason, this effect is neglected here. Vasseur et al. (1983) give an expression for the present temperature disturbance T(z,t=0) in a homogeneous, isotropic half-space whose surface temperature in the past is defined by T(z=0,t), where t is the time before present:

$$T(z, t=0) = \frac{z}{2\sqrt{\pi\kappa}} \int_0^\infty T(z=0, \tau)\tau^{-3/2} \exp\left(\frac{-z^2}{4\pi\tau}\right) d\tau . \tag{12}$$

This expression can be integrated for different functions approximating the variation of the Earth's surface temperature with time. For an abrupt change of ΔT in the Earth's surface temperature at time t equation (12) yields

$$T(z, t=0) = \Delta T \operatorname{erfc}\left(\frac{z}{2\sqrt{\kappa t}}\right) , \tag{13}$$

where erfc is the complementary error function. If the ground temperature history can be approximated by a sequence of K such sudden temperature changes T_k at times t_k, i.e., $T(z=0,t) = T_k$ for $t_{k-1} \le t \le t_k$ (k=1,...,K), then the resulting present temperature disturbance is given by

$$T(z, t=0) = \sum_{k=1}^{K} T_k \left[\operatorname{erfc}\left(\frac{z}{2\sqrt{\kappa t_k}}\right) - \operatorname{erfc}\left(\frac{z}{2\sqrt{\kappa t_{k-1}}}\right) \right] . \tag{14}$$

The inverse problem can now be set up as a system of linear equations:

$$\Theta_j = A_{jk} X_k \quad \text{and} \quad \Theta_j = \Theta(z_j) - M(z_j), \tag{15}$$

where Θ_j is the temperature measured at depths z_j (j=1,...,N), corrected, if applicable, for the input by heat production. The values of X_k are the K+2 unknowns T_0, q_0 and T_1,..., T_K. The matrix A_{jk} then contains ones in the first column and the thermal resistances $R(z_j)$ at the depths z_j in the second column. Columns 3 to K+2 contain the differences between the complementary error functions at the times t_k and t_{k-1}, as defined in equation (14). Thus, a system of N linear equations is defined by equation (15), which must be solved for M=K+2 unknowns. In general, this system of equations is both underdetermined and overdetermined. Moreover, it is unstable. Such systems of equations are often solved by applying the "generalized inverse", e.g., the singular value decomposition (Lanczos 1961; Jackson 1972; Menke 1989). This approach is attractive from a theoretical viewpoint, because the singular value decomposition makes it possible to control the level of noise that is tolerable in the solution. This improvement in the stability of the solution is obtained, however, at the cost of the resolution. In the singular value decomposition the N × M matrix **A** is resolved into a product of three matrixes, $\mathbf{A} = \mathbf{U}\,\mathbf{\Lambda}\,\mathbf{V}^T$, where **U** is an N × N orthonormal matrix in the data space, **V** is an M × M orthonormal matrix in the parameter space and **Λ** is an N × M matrix whose only elements different from zero are the L "singular values" Λ_{ii} (i = 1, ..., L) on the diagonal (L ≤ min(N,M)). The matrixes **U** and **V** consist of the N and M orthonormal column vectors, respectively, which span the data space and the parameter space. If L < M, the system of equations is underdetermined; if L < N, it is overdetermined. A system of equations $\mathbf{Ax} = \mathbf{b}$ is then both underdetermined and overdetermined if, for example, for L < N not all of the equations contain information about all of the unknowns. This is especially the case, for example, in the geothermal inverse problem described above: The number of temperatures measured as a function of depth as a rule by far exceeds the number of temperature changes at the Earth's surface with time to be determined in the inversion. But owing to the controlling transport mechanism, diffusion, which attenuates the contribution of small periods with depth, the temperature field at any depth contains information only for a certain range of periods of surface temperature changes. **Λ** can be partitioned into a submatrix $\mathbf{\Lambda_P}$ of P ≤ L nonzero singular values, where $\mathbf{\Lambda_P}$ is a P × P diagonal matrix, and several zero matrices as

$$\Lambda = \begin{bmatrix} \Lambda_P & 0 \\ 0 & 0 \end{bmatrix}. \tag{16}$$

The decomposition then becomes $\mathbf{A} = \mathbf{U}\,\mathbf{\Lambda}\,\mathbf{V}^T = \mathbf{U_P}\,\mathbf{\Lambda_P}\,\mathbf{V_P}^T$, where $\mathbf{U_P}$ and $\mathbf{V_P}$ consist of the first P columns of **U** and **V**, respectively. The other portions of the eigenvector matrices are cancelled by the zeros in **Λ**. The generalized inverse is given by

$$\mathbf{x} = \mathbf{V}_P\ \mathbf{\Lambda}_P^{-1}\ \mathbf{U}_P^T\ \mathbf{b}\ , \qquad (17)$$

where the M × N matrix $\mathbf{\Lambda}^{-1}$ has L elements $1 / \Lambda_{ii}$ that differ from zero only along its diagonal. These elements are the reciprocals of the "singular values". The instability of the inversion process mentioned above results from the existence of very small singular values whose reciprocals have a large impact on the solution. Noise or errors in the data are amplified by these very small singular values. In practice, there are two ways to avoid this undesired effect: Either only those $P \leq L$ singular values larger than a selected cutoff value are retained, or $\mathbf{\Lambda_P}^{-1}$ in equation (17) is substituted by $\mathbf{\Lambda_P}(\mathbf{\Lambda_P}^2+\varepsilon^2\mathbf{I})^{-1}$, where $\varepsilon > 0$ is an attenuation factor and $\mathbf{I}$ is the unit matrix:

$$\mathbf{x} = \mathbf{V}_P\ \mathbf{\Lambda}_P\ \left(\mathbf{\Lambda}_P^2 + \varepsilon^2\mathbf{I}\right)^{-1}\ \mathbf{U}_P^T\ \mathbf{b}\ . \qquad (18)$$

Large singular values are slightly or not at all affected by either of these two procedures, while, at the same time, they do insure that small singular values do not contaminate the solution. The price for this, however, is a decrease in resolution. The optimum cutoff value or the attenuation factor • must, therefore, be determined in an iterative (and subjective) process which weighs the increase in stability against the loss of resolution. The resolving power of a set of data to identify the wanted parameters of the model (i.e., in our case, the size of the temperature steps at given instants of time) can is determined by the model resolution matrix

$$\mathbf{R} = \mathbf{V}_P\ \mathbf{\Lambda}_P\ \left(\mathbf{\Lambda}_P^2 + \varepsilon^2\mathbf{I}\right)^{-1}\ \mathbf{\Lambda}_P\ \mathbf{V}_P^T\ \ , \qquad (19)$$

where $\mathbf{V}_P$ represents the matrix of the $P \leq L$ vectors in parameter space whose singular values are larger than the given cutoff. The remaining (M - P) columns of the matrix $\mathbf{R}$ are filled with zeros; in this case, $\varepsilon = 0$. If instead of a cutoff value, a value of $\varepsilon > 0$ is used to attenuate the small singular values, then P = L. For a given vector $\mathbf{m}$ of known model parameters, the estimated parameters $\mathbf{m}^{est}$ that would be obtained from the inversion, i.e. the model response, are:

$$\mathbf{m}^{est} = \mathbf{Rm}\ . \qquad (20)$$

The model resolution matrix, therefore, makes it possible to determine whether a data set is suitable for identification of the parameters of the model. If the system of equations is underdetermined ($P \leq L \leq M$), then the solution is ambiguous: Every linear combination of the column vectors of V that is equal to zero or can be neglected can be added to the solution, because these vectors are mapped to the origin of the parameter space. If the system of equations is overdetermined, then the generalized inverse corresponds to the "least squares" inverse (Lanczos 1961). As mentioned above, reconstruction of the ground temperature history from borehole temperature logs is usually both an over- and underdetermined problem.

Because long-term meteorological trends correlate over distances distinctly longer than 500 km (Hansen and Lebedeff 1987; Schönwiese et al. 1994), temperature logs in different boreholes in the same region may be influenced by the same ground temperature history. This assumes the same surface conditions for all boreholes, e.g vegetation, insolation, precipitation, etc.. If this is the case, then the temperature disturbances for these boreholes should be the same. If it can be assumed that errors and noise in data from different boreholes do not correlate with each other, then stacking of the temperature disturbances derived from different boreholes should improve the signal-to-noise ratio. The stacking involves only the temperature disturbances and not the measured temperatures, because the surface temperature and the heat flow density can be different at every borehole. In practice, the temperature data for all boreholes in a region are inverted simultaneously in the stacking procedure. For I boreholes, therefore, the unknowns are the I surface temperatures, the I heat flow density values, and the K model parameters for the ground temperature history. The data are the temperature logs from all of the boreholes. The structure of the matrix for the inversion of these I boreholes is similar to that for a single borehole: If N_i is the number of temperature measurements in the i-th borehole, then the matrix has $N_1+N_2+\ldots+N_I$ rows and K+2I columns. The first N_1 elements of the first column are ones and all others are zero. The following N_2 elements of the second column are also ones and all others are zero, etc. The next I columns are obtained by replacing all ones in the previous I columns by the thermal resistances at depths $z_j^{(i)}$ ($z_j^{(i)}$: depth z_j in the i-th borehole). The remaining K columns are obtained by subtracting the complementary error function at time t_k from that at time t_{k-1} for each depth in each borehole:

$$\begin{bmatrix} \boldsymbol{\Theta}^{(1)} \\ \boldsymbol{\Theta}^{(2)} \\ \cdots \\ \boldsymbol{\Theta}^{(I)} \end{bmatrix} = \begin{bmatrix} 1\ 0 \ldots 0 & \mathbf{R}^{(1)} & 0 & \ldots & 0 & \mathbf{A}^{(1)} \\ 0\ 1 \ldots 0 & 0 & \mathbf{R}^{(2)} & \ldots & 0 & \mathbf{A}^{(2)} \\ \cdots\cdots & \cdots & \cdots & \cdots & \cdots & \cdots \\ 0\ 0 \ldots 1 & 0 & 0 & \ldots & \mathbf{R}^{(I)} & \mathbf{A}^{(I)} \end{bmatrix} \begin{bmatrix} \mathbf{T}_0^{(1)} \\ \mathbf{T}_0^{(2)} \\ \cdots \\ \mathbf{T}_0^{(I)} \\ \mathbf{q}_0^{(1)} \\ \mathbf{q}_0^{(2)} \\ \cdots \\ \mathbf{q}_0^{(I)} \\ \mathbf{T} \end{bmatrix}, \tag{21}$$

where $\mathbf{\Theta}^{(i)}$ is the vector of the temperature values for the i-th borehole, $\mathbf{R}^{(i)}$ is the vector of the thermal resistances at each depth in the i-th borehole, and $\mathbf{A}^{(i)}$ is the matrix formed according to equation (14) by subtracting the complementary error function at time t_k from that at time t_{k-1} for each depth in the i-th borehole. The unknown parameters are the I reference temperatures $T_0^{(i)}$ at the Earth's surface, the I reference heat flow densities $q_0^{(i)}$, and the K model parameters of the ground temperature history (i.e. the size of the temperature steps at given instants of time) in the vector **T**. It is assumed that all of the boreholes document the same information, i.e., the temperature logs must all cover the same depth range and have the same resolution. This procedure was successfully used in eastern Canada, where the temperature logs of numerous boreholes record consistent signals (Beltrami and Mareschal 1992). Its use is illustrated in the following discussion using a data set from 15 shallow boreholes in Bavaria and neighboring Bohemia.

2.2.1.1 The Temperatures of the Last 200 Years at the Earth's Surface in Central Europe from Temperature Logs from Shallow Boreholes in Bohemia and Bavaria

The 15 boreholes discussed here are along the western margin of the Bohemian Massif, a high-grade metamorphic Variscan crystalline basement complex. Nine of these shallow boreholes are in the immediate vicinity of the KTB borehole in northeastern Bavaria, another six are just over the border in Bohemia. Their names and symbols are given in Table 2.1; their locations are shown in Figure 2.8. Their coordinates, elevations, depths, and temperature at a depth of 100 m are given in Table 3.1 (section 3.1.1). Most of the boreholes in Bavaria were drilled within the scope of the geophysical site survey of the German Continental Deep Drilling Project (KTB) (Burkhardt et al. 1989). The boreholes in Bohemia were drilled for mineral exploration. Most of the boreholes are within a radius of 50 km. In the northern part of the area, the Bavarian and Bohemian boreholes are separated by the Bavarian and Bohemian Forests (Oberpfälzer Wald and Česky Les, respectively, with elevations up to 800 m) and in the southern part of the area by the Šumava Mts. (with elevations up to 1300 m) (Fig. 2.8). Where available, mean values for thermal conductivity, thermal diffusivity, and heat production rate for the boreholes are also given in Table 2.1. The data were taken from Burkhardt et al. (1989) and Čermák (1975, 1981). For the inversion, values were picked at intervals of about 3 m from the continuous temperature logs recorded in the German boreholes. In the Czech boreholes temperature had been recorded at 10 m interval. Plate 4 shows the 15 temperature logs in reduced form. Further information about the data and their analysis is given in Clauser and Mareschal (1995).

The first attempts to invert individual temperature logs from the boreholes of the KTB project did not yield consistent results. This is probably due to small hydraulic disturbances and other noise that may not be a problem when averaging techniques, for example, Bullard plots, are used to determine heat flow density (cf. Burkhardt et al. 1989). In contrast, the results of an inversion are very sensitive to such noise, because the inversion procedure attempts to fit all deviations from a linear temperature gradient by an appropriate ground temperature history. Thus, these temperature disturbances in a borehole consist of a real climatic signal together with noise from numerous other sources, for example, heterogeneity and anisotropy of the thermal conductivity (which cause lateral refraction of the otherwise vertical heat flow), groundwater flow, etc. However, because the boreholes penetrate different geological units, and owing to the faulting, folding and foliation of the crystalline basement of this region, it is very unlikely that these disturbances correlate between boreholes. Under these conditions, and if there is a sufficient number of boreholes in a study area, stacking of the information from these temperature logs will improve of the signal-to-noise ratio: If the ground temperature histories at the locations of the boreholes are similar, then joint inversion of the data from these boreholes will attenuate the influence of individual disturbances and yield only that signal which is common to all of the borehole sites. The reference heat flow density q_0 for the individual boreholes resulting from the inversion of the three data sets (Bavaria, Bohemia, Bavaria and Bohemia is given in Table 2.1.

In section 3.1.1 vertical Péclet numbers and groundwater recharge rates are discussed which were determined from those nine Bavarian temperature logs used here to reconstruct the ground temperature history. The results are ambiguous, and the attempt to fit a lateral trend to the Péclet numbers was only partially successful. Thus, it is expected that reconstruction of a climate signal using the "stacking" property of the joint inversion can make it easier to distinguish between climatic and hydrological causes of the curvature of temperature logs.

The inversion process presented here does not lend itself to automatization, because the number of singular values used for the inversion is determined by the selected cutoff value. It determines the relationship between the resolution of the model parameters (mainly the surface paleo-temperatures) and the stability of the solution. Therefore, the inversion is begun with a very small cutoff value, e.g., $10^{-3} - 10^{-4}$. The result of such an inversion is normally a strongly fluctuating ground temperature history, because the small singular values strongly amplify the effect of noise. The cutoff value is then increased step by step, eliminating further singular values. After several iterations, the unphysical oscillations, which are typical of small cutoff values, disappear from the ground temperature history determined in this way. However, if too many singular values are excluded from the solution by using cutoff values that are too large, then the resolution will be so low that the inversion no longer yields any information about the ground temperature history. It is apparent that this procedure is ambiguous and depends on the judgment of the person who is analyzing the data. However, additional, independent information can be used to support this process. For example, the results of this study can be compared with mean annual air temperature at the five meteorological stations Berlin, Jena, Bayreuth, Prague, and Munich. Different cutoff values were used for the three data sets, but a cutoff value of 0.05 yielded the best results. The reasons for this are discussed in Clauser and Mareschal (1995).

Table 2.1. Name, number, year of the temperature log, mean values for thermal conductivity λ, thermal diffusivity κ and heat production rate H for the 15 borehole logs used for the joint inversion (Plate 4), as well as reference heat flow densities q_0 from the inversion of the data from the Bavarian and Bohemian boreholes and the two groups of boreholes together (from left to right). The number of measurements is given in parentheses. Values for Falkenberg are mean values for two neighboring boreholes. A further thermal diffusivity value is available for the Poppenreuth borehole ($\kappa = 0.926$ mm^2 s^{-1}), which was not included in the inversion owing to hydraulic disturbances. The mean thermal diffusivity for the eight localities in Bavaria is $\kappa = 0.84 \pm 0.06$ mm^2 s^{-1}. Thermal diffusivity was not determined for the Bohemian boreholes and values for heat production rate are only available for some of them.

borehole	No.	year logged	λ [W m^{-1} K^{-1}]	κ [mm^2 s^{-1}]	H [μW m^{-3}]	q_0 [mW m^{-2}]	q_0 [mW m^{-2}]	q_0 [mW m^{-2}]
Wülfersreuth	1	1986	3.60 (12)	0.891 (1)	2.33 (9)	66.2	–	66.3
Weissenstadt	2	1986	3.42 (16)	0.900 (1)	9.80 (16)	77.0	–	77.2
Röthenbach	3	1986	2.64 (15)	0.820 (1)	3.46 (20)	50.8	–	49.9
Neusorg	4	1986	3.77 (24)	0.778 (1)	2.69 (17)	79.5	–	80.5
Falkenberg NB3, PB7	5	1984	3.45 (20)	–	6.74 (15)	78.8	–	79.7
Püllersreuth	6	1986	3.02 (42)	0.785 (1)	1.84 (17)	64.6	–	64.3
Remmelberg	7	1986	2.73 (14)	0.767 (1)	0.19 (20)	74.9	–	74.0
Griesbach	8	1986	3.43 (74)	0.832 (1)	2.10 (20)	77.2	–	78.0
Zadni Chodov	A	1975	2.34 (7)	–	1.26 (?)	–	62.4	63.4
Vitkov	B	1972	2.40 (39)	–	3.14 (41)	–	61.8	60.6
Pernolec	C	1987	2.40*	–	–	–	61.8	62.6
Lom u Stribra	D	1977	4.09 (12)	–	–	–	64.7	63.9
Bor u Tachova	E	1975	2.71 (11)	–	–	–	54.2	54.0
Holubov	F	1972	2.89 (16)	–	0.63 (30)	–	53.6	53.2

* Because it is so close to the Vitkov borehole, the same thermal conductivity value was assumed for the Pernolec borehole.

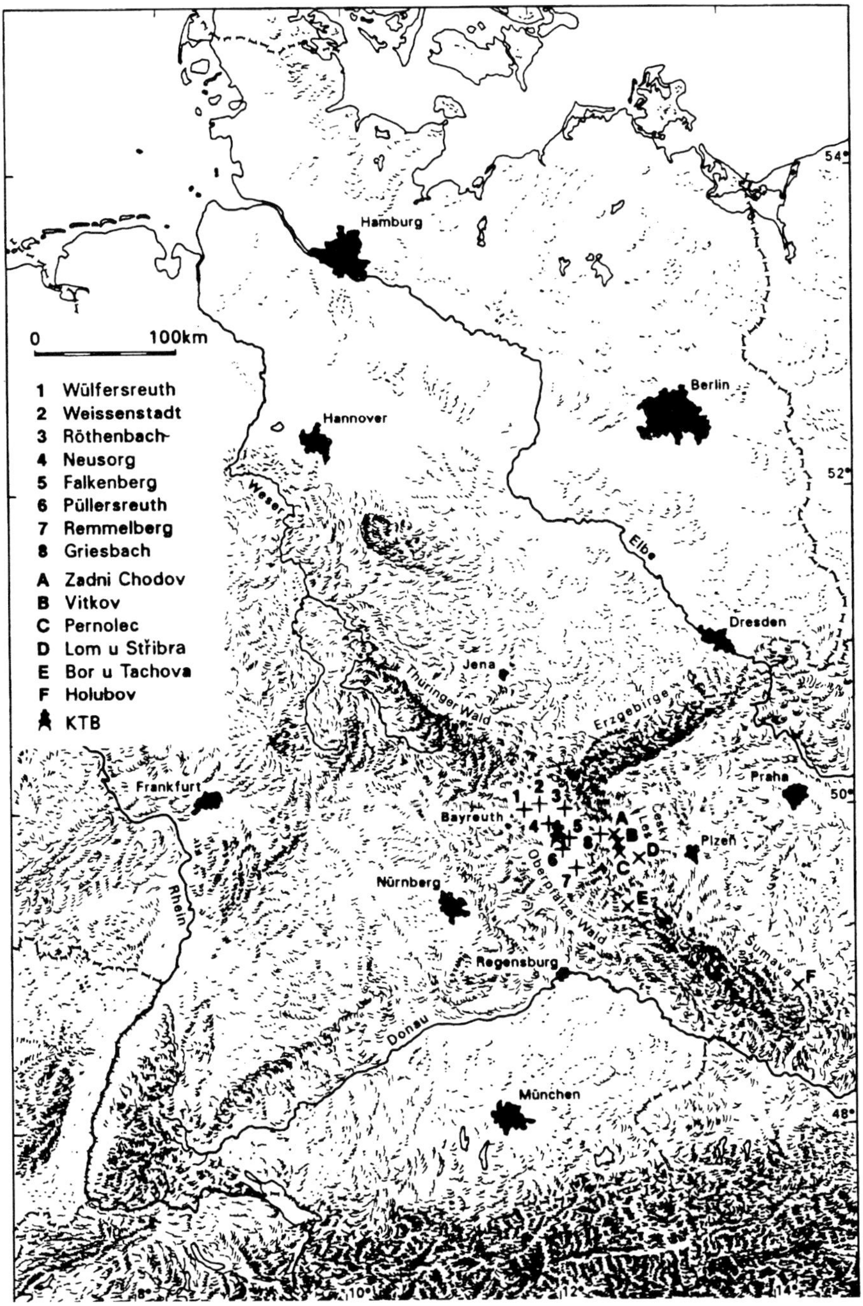

Fig. 2.8. Locations of the shallow boreholes used for this study (Table 2.1). Long-term records of air temperature are available from meteorological stations at Berlin, Jena, Bayreuth, Prague (Praha) and Munich (München).

Number and length of the time intervals of the ground temperature history to be inverted must also be determined. In general, it is preferable to use too many parameters for the model, because then the linear combination of parameters that is the most strongly constrained by the data is selected by the singular value decomposition procedure. In spite of this, optimization of the procedure could certainly save computation time and improve the stability of the solution.

The results of the inversion shown here are for an analyzed interval of 300 years subdivided into 30 intervals of 10 years each. Because the thermal diffusivity varied little relative to the thermal conductivity (Table 2.1), thermal diffusivity values for the Bohemian boreholes were not calculated from the thermal conductivity values and an assumed value for the thermal capacity ρc. Instead, a mean thermal diffusivity of $\kappa = 0.84 \text{ mm}^2 \text{ s}^{-1}$ (or $\kappa = 26.5 \text{ m}^2 \text{ a}^{-1}$) was used for all data sets. These values were determined on cores from 10 boreholes in Bavaria (one borehole, Poppenreuth, was not included in the inversion owing to large hydraulic disturbances). The heat production rates measured on cores from the Bavarian and Bohemian boreholes ranged from $0 - 10\ \mu\text{W m}^{-3}$. Because a high heat production rate influences the curvature of temperature logs from shallow boreholes, those temperature logs for which values for the heat production rate are available (Table 2.1) were reduced by the amount of this effect. In spite of a high heat production rate in some boreholes, the correction was smaller (< 100 mK) than the temperature disturbance (500 mK) resulting from the ground temperature history.

The results of the joint inversion can be compared with the data from each of the boreholes. This is illustrated in Plate 5 for the inversion of the data from the nine Bavarian boreholes and the Falkenberg PB7 borehole. The measured T log (blue circles) is shown in the upper left, together with the temperature profile predicted by the model (red line) and the reference profile prior to the transient temperature perturbation (dashed green line). Because the difference between the model prediction and the data cannot be readily seen at this scale, it is plotted separately in the upper right. The difference is small (< 50 mK), but not insignificant: Its size and shape are determined by that eigenvector in data space that belongs to the smallest singular value used in the inversion. The small superposed fluctuations of about 5 mK amplitude are due to other disturbances (heterogeneity, anisotropy, groundwater, etc.) and noise. The two lower graphs show the perturbations in temperature and heat flow density, which are related to the ground temperature history. These signals are calculated from the difference between the reference or equilibrium conditions and the observed profiles. In our case, the magnitude of these signals below a depth of 150 m is less than the accuracy of the measured temperature and heat flow density data.

The model explains the data well, as can be seen in Plate 5. This result is found in many inversion studies of temperature logs and cannot be due to only the effect of using too many parameters. Using a cutoff value of 0.05, leaves 23 singular values for the inversion. The eigenvectors in the data space corresponding to the 18 largest singular values represent, in principle, the nine T_0 and q_0 values from each borehole, i.e., the reference values for the temperature at the Earth's surface and the heat flow density. They contain only very small components of the ground temperature history. The values of the model resolution matrix **R**, which is determined from these vectors, are very close to one along the matrix diagonal and otherwise are equal to zero. The other five vectors contain mainly the components of the ground temperature history. For these vectors, the values along the diagonal are no longer significantly larger than the off-diagonal values, however. Thus, the data can be used to resolve only the long-period components of the ground temperature history, and short-period fluctuations are lost in the inversion process. This is in keeping with the character of the diffusion process, which strongly attenuates the higher frequency signal components with increasing depth. As a consequence, essentially only long-period components are present at depth.

The ground temperature history obtained by inversion of the data from the nine Bavarian temperature logs is shown in Figure 2.9. The error envelope is a measure of the propagation of an assumed measurement error of 100 mK in the inversion process. It does not give the uncertainty in the ground temperature history as a function of time, but rather the uncertainty in a mean value for the interval defined by the structure of smallest eigenvector used in the inversion. The main trends of this ground temperature history are: (1) cooling from 1750 to 1900, (2) warming from 1900 to 1950, and then (3) a series of cooling, warming and cooling periods.

Studies using synthetic, noise-free data have shown that when the number of singular values used in the calculations is too small, the inversion can lead to artificial oscillations in the ground temperature history (Mareschal and Beltrami 1992). Comparison with meteorological data seems to confirm, however, the frequency of the oscillations in this simulated ground temperature history after 1950 (Fig. 2.9), although not to the same extent as their amplitudes.

The fit of the model to the data from the six Bohemian temperature logs was as good as that for the Bavarian data. A cutoff value of 0.05 yields 18 singular values for the calculations. Twelve of these contain primarily the information about the values for T_0 and q_0 at the six boreholes. Thus, there are six singular values that contain mainly the components of the ground temperature history, one more than for the Bavarian data. Inversion of the combined Bavarian and Bohemian data sets yields 36 singular values larger than the cutoff value of 0.05. Thirty of these represent mainly T_0 and q_0 at the 15 boreholes; therefore, also in this case, six singular values characterize the ground temperature history.

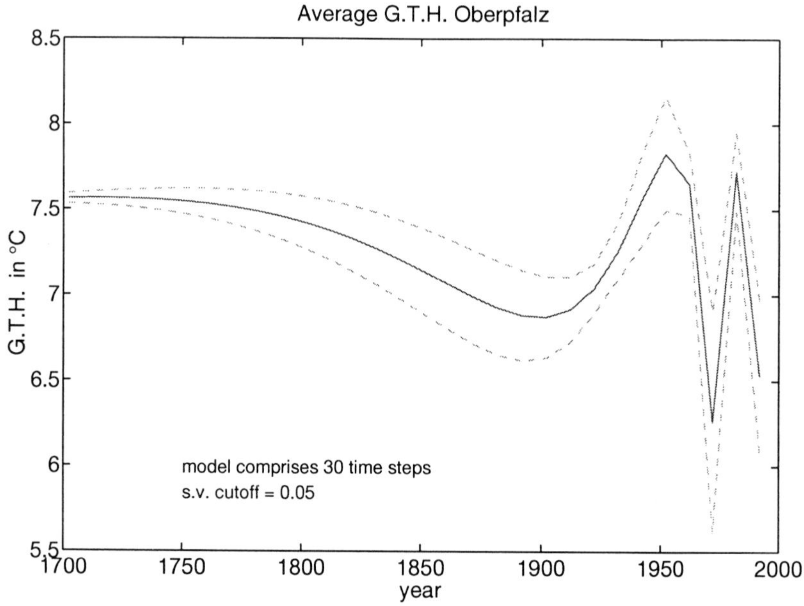

Fig. 2.9. Paleo-temperature at the Earth's surface, reconstructed by inversion of the data from nine temperature logs from boreholes in Bavaria: The error envelope is a measure of the variability of a long-period mean or of the propagation of an assumed measurement error of 0.1 K in the inversion – see text and Table 2.1 for further details and Fig. 2.8 for the locations of the boreholes.

All three ground temperature histories obtained by inversion of the three data sets are shown in Plate 6. They are not identical: In contrast to the ground temperature history of Bavaria, that of Bohemia shows a slight warming from about 1700 to 1840, followed by a period of rapid cooling until about 1930. The fluctuations that have occurred since then are smaller than in Bavaria, and instead of the present cooling there, the warming trend that began about 1960 has continued. Because all except one of the Bohemian temperature logs were measured before 1977, the recent climate history derived from them cannot be expected to be as good as in the temperature logs recorded later in the Bavarian boreholes. Inversion of the combined data sets yields a ground temperature history more or less between those in Bavaria and Bohemia. The main trends are a cooling of 0.85 K from 1800 to 1920, followed by more recent fluctuations, as in Bavaria, but with slightly elevated amplitudes with a maximum of ± 1 K. Comparison of the error envelope of the three ground temperature histories shows no significant improvement in the case of the combined data sets. Like for the other two ground temperature histories, the standard deviations (0.2 – 0.5 K) of the fluctuations of the last five decades are rather large. Therefore, fluctuations with smaller amplitudes cannot be resolved within these limits of error.

The large recent fluctuations in the reconstructed ground temperature histories require verification by comparison with other data. Fortunately, air temperature data is available from the meteorological stations at Berlin, Jena, Bayreuth, Prague, and Munich, some of which extend back into the 18th century. Origin and consistency of these data sets are discussed by Clauser and Mareschal (1995). For comparison with the three reconstructions of ground temperature histories discussed above, moving averages were calculated for 5- or 10-year intervals from these mean annual temperatures. These are also shown in Plate 6. The short-period peaks of these five sets of data correlate very well over a period of more than 200 years. This is especially the case for the long-period trends in the records from Prague, Jena, Munich, and Bayreuth. Although these cities are within 160 km of the study area, they are separated by mountains several hundred meters high (the Thuringian, Bavarian, and Bohemian Forests, see Fig. 2.8). In contrast, the long-period trends of the data from Berlin are phase-shifted relative to the borehole data. Berlin is the furthest from the study area, and the data may be influenced by its nearness to the Baltic Sea, as well as by the "heat island" effect of the metropolitan area.

Comparison of the three ground temperature histories with the meteorological records shows reasonable agreement: (1) The meteorological data for all of the stations show a minimum between 1830 and 1890. This is in good agreement with the Bavarian ground temperature history. The poorest agreement is with the Bohemian ground temperature history, whose minimum occurs between 1930 and 1940. The early warming in this ground temperature history is perhaps best found in the data from Berlin and Prague, whereas the data from Jena, Bayreuth and Munich correlate better with the Bavarian ground temperature history. On the whole, the agreement of the early part of the ground temperature history for Bohemia with the meteorological data does not appear to be as good as that of the one for Bavaria. The ground temperature history of the combined data set lies between those of the two areas. (2) The short-period fluctuations of the last 50 years are found in all of the ground temperature histories, sometimes with a slight time shift. (3) After 1970 the data from Jena and Munich, as well as the Bohemian ground temperature history, show a continued warming. In contrast, following the maximum in 1970 the stations Prague, Berlin, Bayreuth (the station closest to the study area), and the ground temperature histories of Bavaria and of the combined data set all show another minimum between 1975 – 1982. After 1982, the temperature increases again. This last period in the ground temperature histories cannot be resolved owing to the year in which the temperature logs were recorded (Table 2.1) and the length of the time intervals (10 years) used for the inversion. Summarizing, the most important properties of the reconstructed ground temperature histories appear to be reflected well in the variations in the long-term records of air temperature in the surrounding meteorological stations.

Another possibility for comparing meteorological records and inversion results is provided by equation (20). Multiplication of the vector of a meteorological time series **m** and the model resolution matrix **R** yields the vector $\mathbf{m}^{est}$ of model parameters that would be estimated by inversion. A ten-year mean of the air temperature, corresponding to the length of the time intervals used for the inversion, was used for this comparison. Additionally, the resulting time series was reduced by its mean and, if necessary, filled with zeros until its length corresponded to the number of columns of the model resolution matrix. The model input **m** is compared in Plate 7 with the model response $\mathbf{m}^{est}$ for the combined Bavarian and Bohemian data set and the corresponding ground temperature history.

This comparison was made with the model resolution matrices of all three data sets. Several conclusions can be drawn from the results of these comparisons: (1) The only model response that begins with a minimum is that for the Berlin time series. (2) Because the locations of the boreholes are not the same as those of the weather stations, and because the soil temperature is influenced by a number of other factors (see, for example, Kappelmeyer and Hänel 1974, Beltrami and Mareschal 1991), this comparison can be only qualitative. Nevertheless, the trends are similar, even if the ΔT amplitudes of the ground temperature histories and the model responses are different. The ground temperature history reconstructed from the combined data set correlates best with the model responses for the data sets from Prague, Jena, and Munich. For the Berlin temperature data, the agreement is less satisfactory, with the exception of the last 50 years. (3) The Bohemian ground temperature history correlates satisfactorily with the model responses only for the last 50 years. (4) Thus, the ground temperature history of the combined data set shows, in general, a good correlation with the model responses, although perhaps in some parts not quite as well as those for the Bavarian data.

Although the reconstructed ground temperature histories for the last 250 years are generally in good agreement with the meteorological records, there are distinct differences from other paleo-climatic histories reconstructed from temperature logs. Šafanda and Kubik (1992) studied the temperature logs from two boreholes (700 m and 2000 m deep) in neighboring Bohemia. Owing to their greater depth, these boreholes provide temperature information further into the past than the shallow boreholes discussed above. Moreover, the inversion method used by Šafanda and Kubik (1992) yields only a single temperature jump instead of a complete ground temperature history. Inversion of the data from the deeper of the two temperature logs indicates a warming of 6 K at the end of the Würm Ice Age about 12,000 years ago. The shallower T log shows a warming of 1 – 2 K 480 years ago, which appears to be associated with local clearing of forests. This event is, therefore, not reflected in the regional ground temperature history. Because these two events observed by Šafanda and Kubik are not recognizable in the regional ground temperature histories, no further comparison is possible. Vasseur et al. (1983) and Mareschal and Vasseur (1992) determined a ground

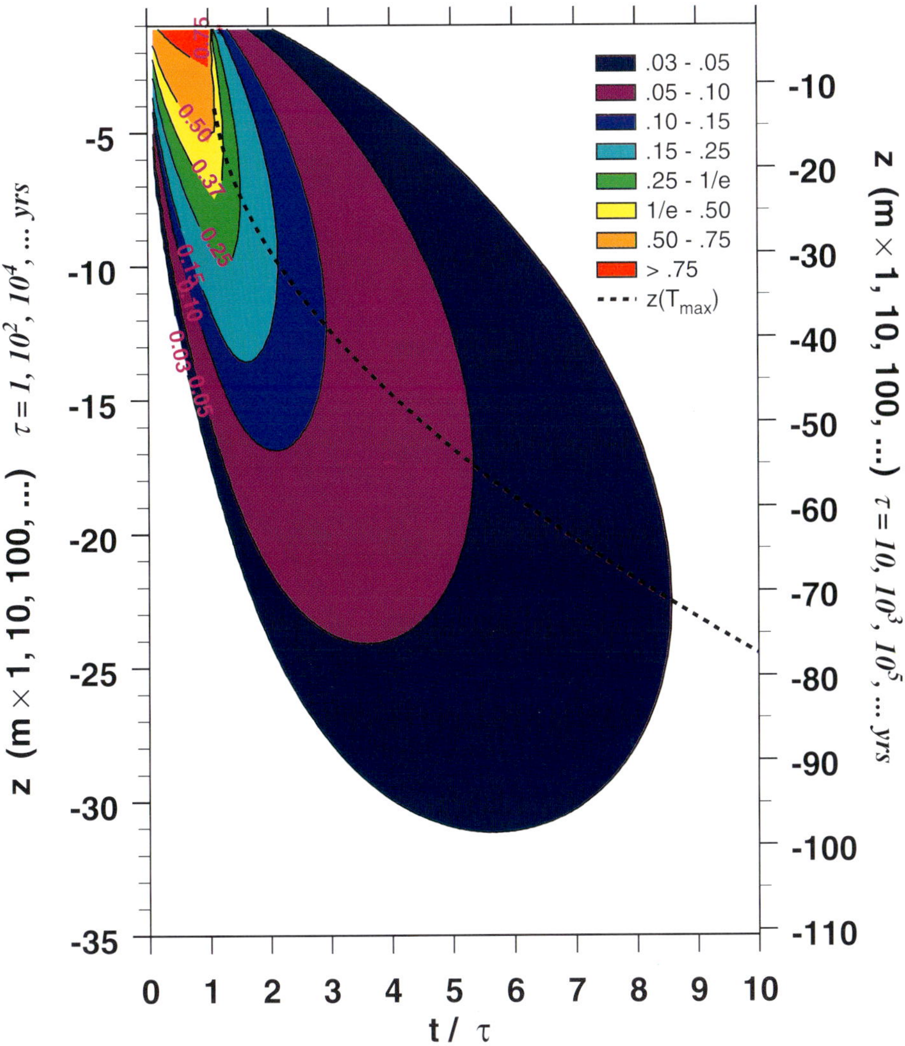

Plate 1. Diffusion of a rectangular temperature signal ΔT of duration τ with time t and depth z from the surface into an isothermal half-space of constant thermal diffusivity $\kappa = 1\ mm^2\ s^{-1}$ (isolines normalized with respect to ΔT). The dashed line marks the depth of the maximum of the diffused temperature signal as a function of time. The depth axis can be normalized with respect to any signal duration τ by multiplication by $\tau^{1/2}$, as is demonstrated on the left and right depth axes for τ = 1, 100, 10,000, ... and τ = 10, 1000, 100,000, ... years, respectively.

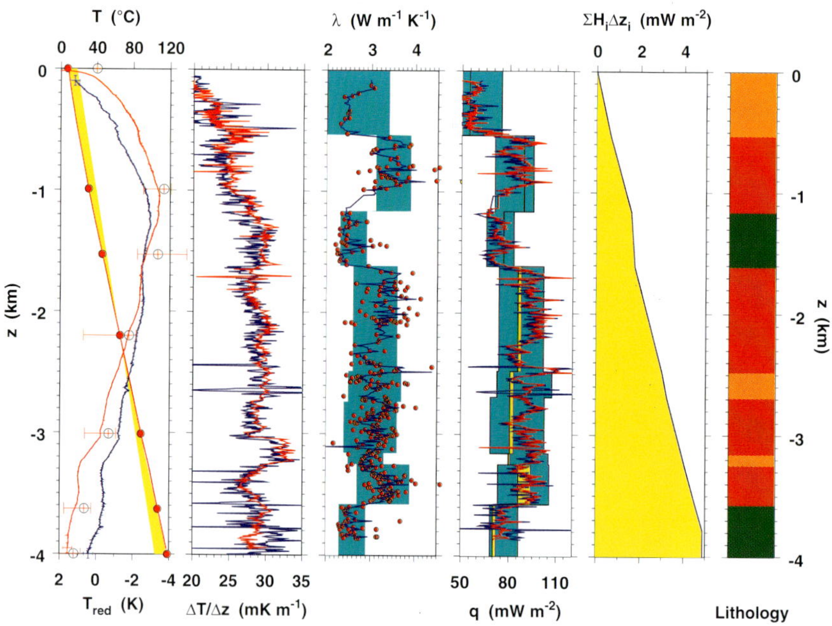

Plate 2. Logs of thermal data from the KTB pilot borehole (see p.17 for details).

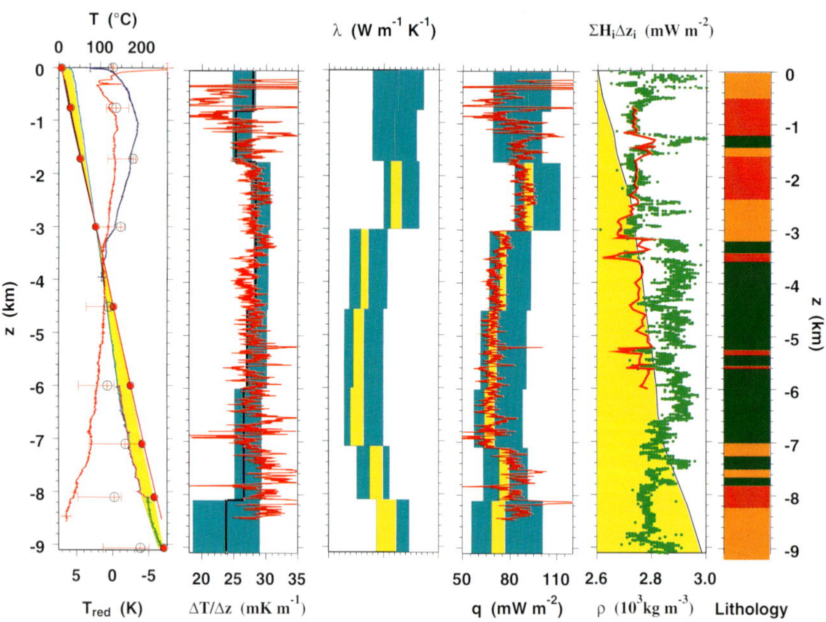

Plate 3. Logs of thermal data from the KTB main borehole (see p.18 for details).

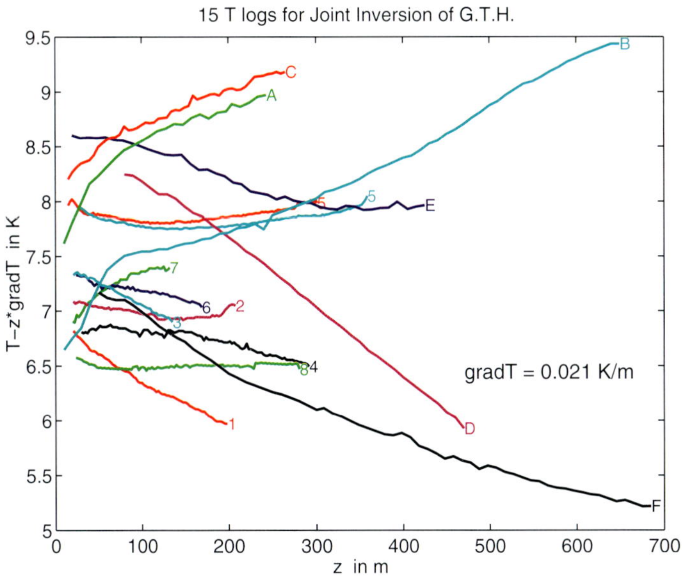

Plate 4. Residual temperature in 15 boreholes used for joint inversion: All of the temperature logs were reduced by a constant temperature gradient of 21 mK m^{-1}. The boreholes are labeled as in Table 2.1. At location 5 (Falkenberg) there are two boreholes within 300 m of each other.

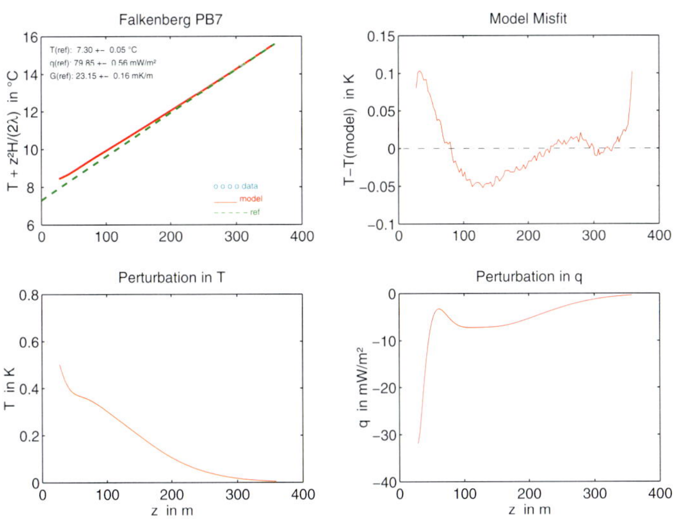

Plate 5. Results of the joint inversion of the 15 Bavarian and Bohemian temperature logs (see Table 2.1 and Fig. 2.8) using 0.05 as singular value cutoff; plottted are fit of model to data, and climate-related perturbations in temperature and heat flow density for borehole Falkenberg PB7. T(ref), q(ref), and G(ref) are the reference values for the steady-state temperature at the Earth's surface, and for undisturbed heat flow density and temperature gradient.

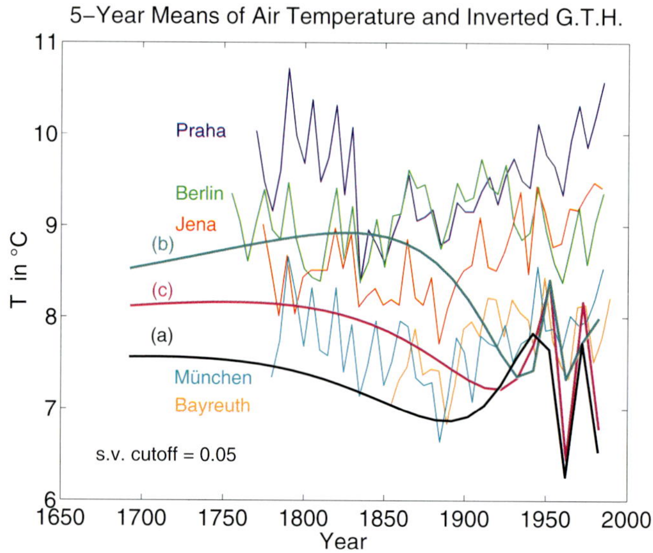

Plate 6. Five-year means of the air temperature at Prague, Berlin, Jena, Munich and Bayreuth (from top to bottom) and the ground temperature history obtained by inversion of the (a) Bavarian, (b) Bohemian and (c) combined data sets.

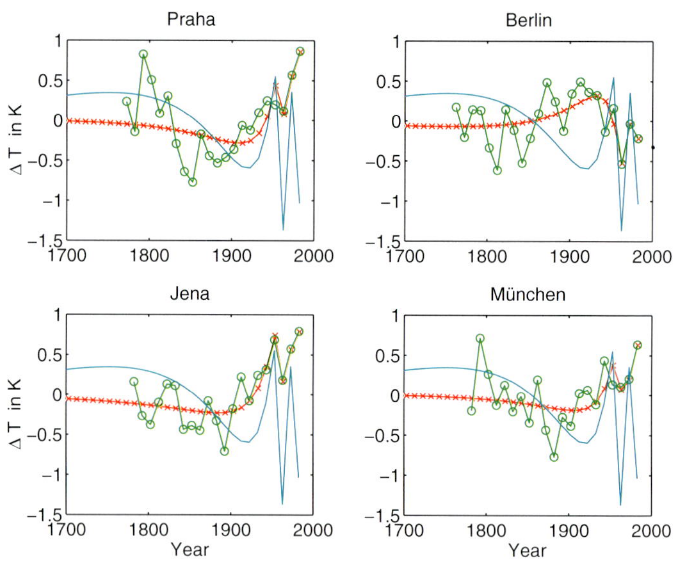

Plate 7. Model response $\mathbf{m}^{est}$ (, equation 20) to 10-year means of air temperature (○) in Prague, Berlin, Jena and Munich: The solid, cyan line is the ground temperature history obtained by inversion of the combined data set from nine Bavarian and six Bohemian temperature logs.

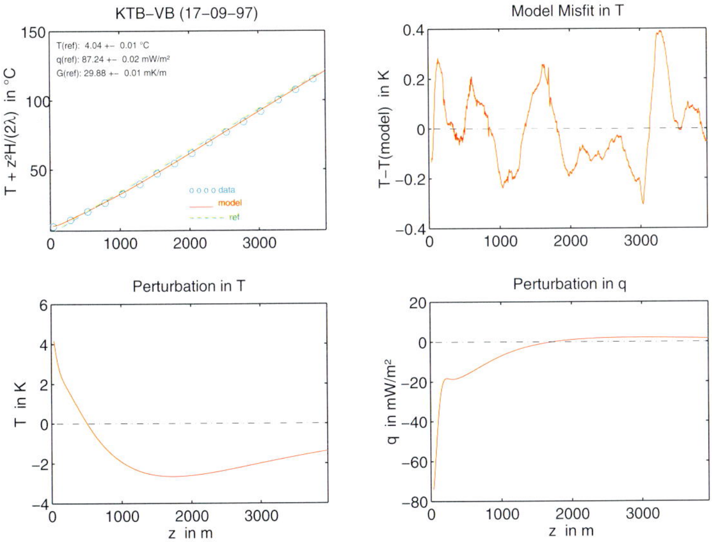

Plate 8. Results of inversion of the KTB pilot borehole temperature log using an attenuation factor of ε=0.2; plottted are fit of model to data, and climate-related perturbations in temperature and heat flow density. T(ref), q(ref), and G(ref) are the reference values for the steady-state temperature at the Earth's surface, and for undisturbed heat flow density and temperature gradient.

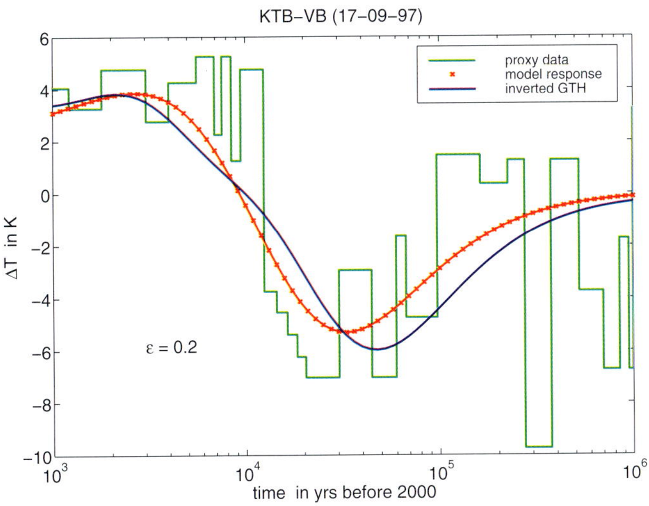

Plate 9. Reconstruction of the ground temperature history by inversion (using an attenuation factor of ε=0.2) of the temperature log recorded in the KTB pilot borehole on 17 Sept. 1997 (Plate 2); the model response (equation 20) to a cubic spline interpolation of the proxy curve of paleo-temperature for Germany (Zoth and Hänel, 1988) is shown for comparison. In each case, the difference from the present conditions is plotted.

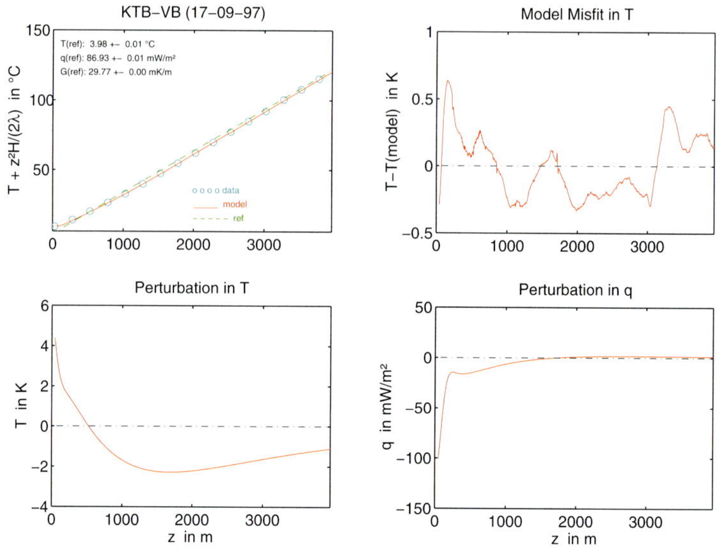

Plate 10. Results of inversion of the KTB pilot borehole temperature logs using an attenuation factor of ε=0.4; plottted are fit of model to data, and climate-related perturbations in temperature and heat flow density. T(ref), q(ref), and G(ref) are the reference values for the steady-state temperature at the Earth's surface, and for undisturbed heat flow density and temperature gradient.

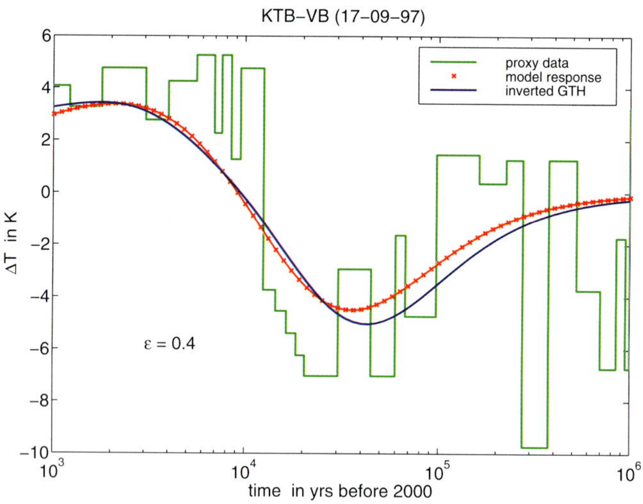

Plate 11. Reconstruction of the ground temperature history by inversion (using an attenuation factor of $\varepsilon = 0.4$) of the temperature log recorded in the KTB pilot borehole on 17 Sept. 1997 (Plate 2); the model response (equation 20) to a cubic spline interpolation of the proxy curve of paleo-temperature for Germany (Zoth and Hänel, 1988) is shown for comparison. In each case, the difference from the present conditions is plotted.

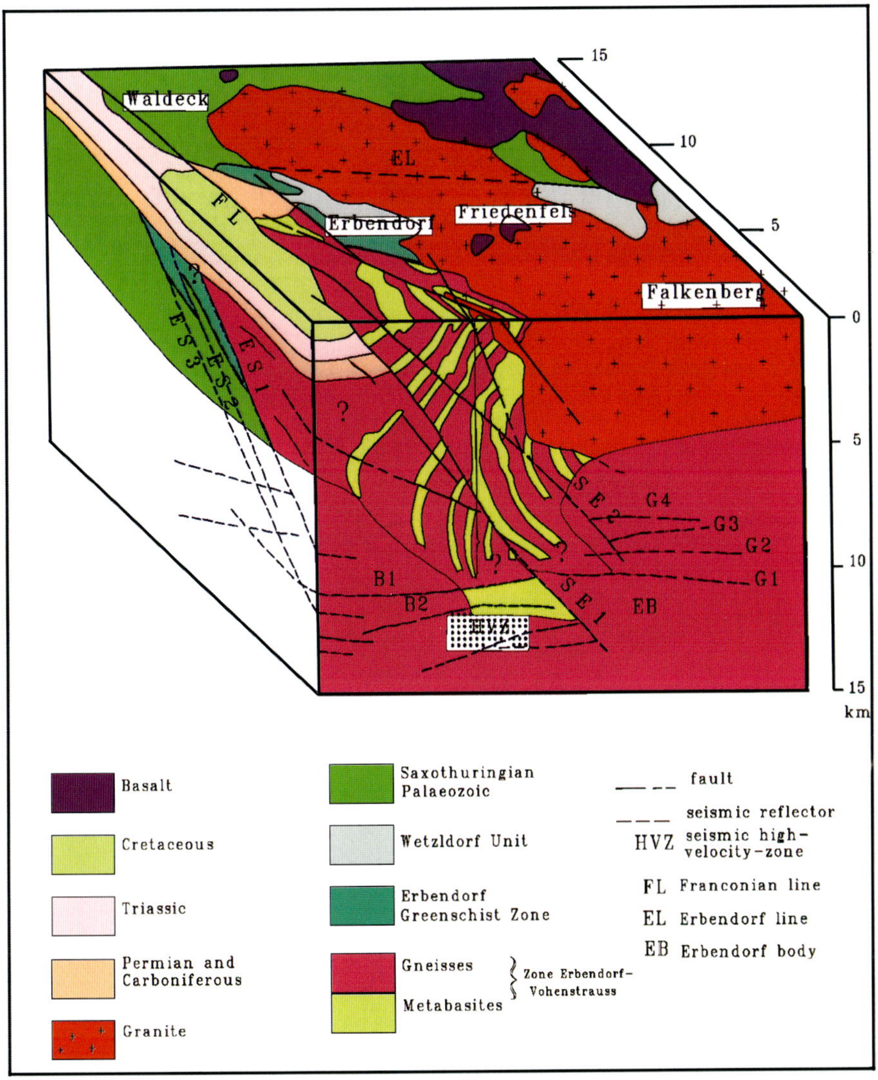

Plate 12. Geological structure in an E–W section through the KTB site, showing geological units, lithological structures, and seismic reflectors (B, ES, G, SE) according to the results of 3-D reflection seismics (after Hirschmann 1993).

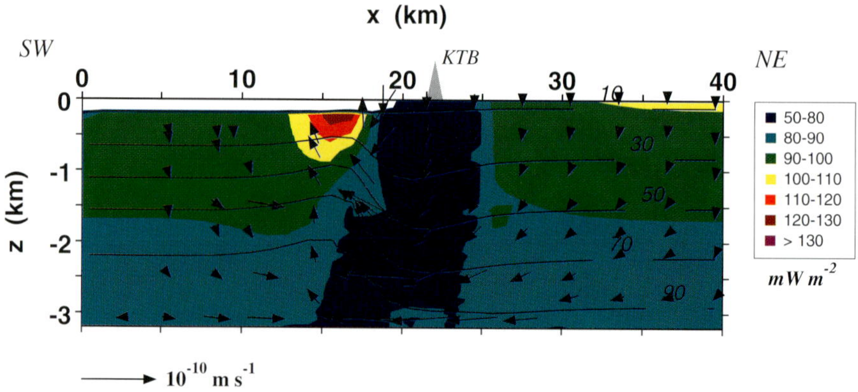

Plate 13. Temperature (isolines in °C), heat flow density (color) and Darcy velocity (arrows) in the model for shallow depths (Fig. 3.8): Doubling of the length of an arrow corresponds to a tenfold velocity. The permeability distribution used for this model is described in the text (for other parameter values, see Table 3.3).

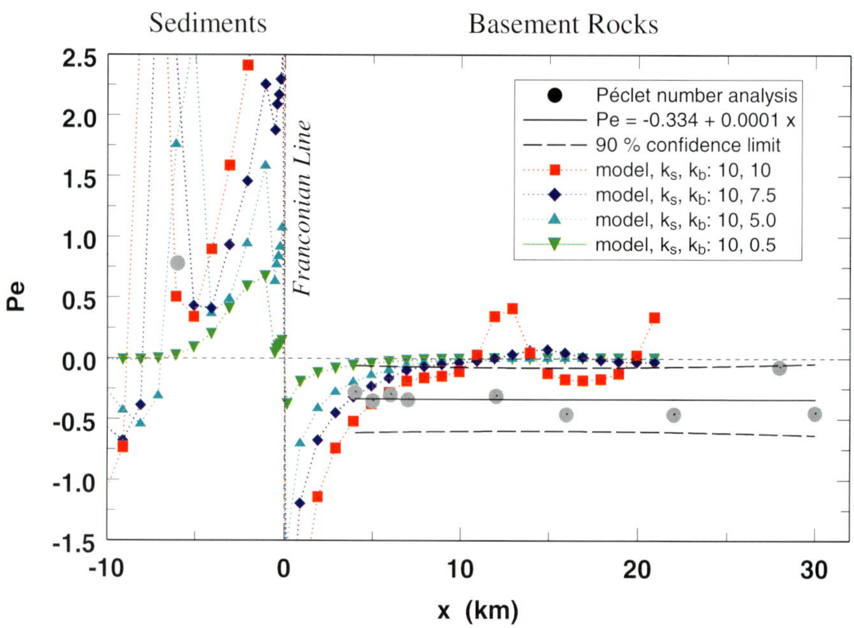

Plate 14. Comparison of the results of the model for shallow depths with those of the Péclet number analyses, projected onto profile AB perpendicular to the Frankonian Line (Fig. 3.2): From left to right, the boreholes are in the same order as in Table 3.2. The permeabilities of the sedimentary and the crystalline basement units, k_s and k_b, are given in 10^{-15} m^2.

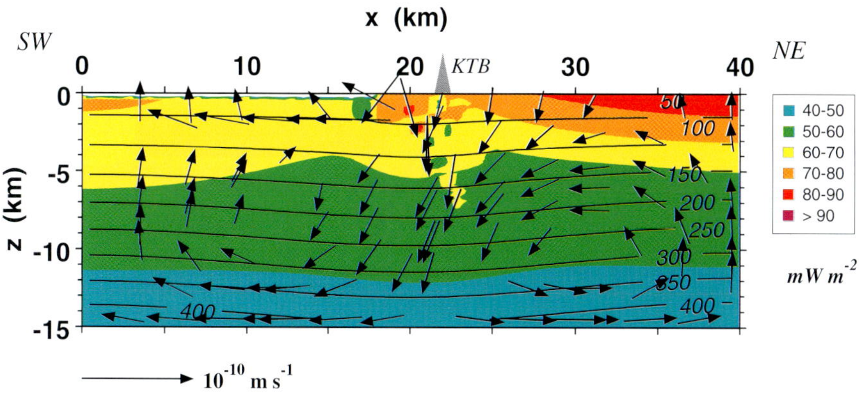

Plate 15. Temperature (isolines in °C), heat flow density (color) and Darcy velocity (arrows) in the model for the upper crust (Fig. 3.17): Doubling of the length of an arrow corresponds to a tenfold increase in velocity. The permeability of the mid-crust is 2.5×10^{-17} m^2 (see Table 3.5 for further parameter values).

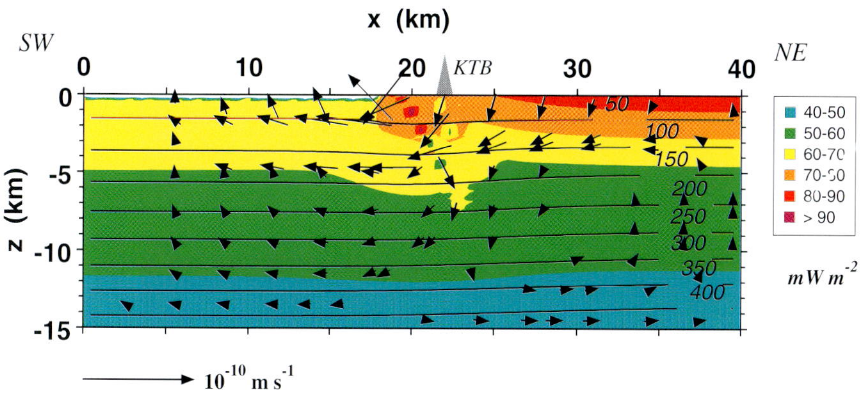

Plate 16. Temperature (isolines in °C), heat flow density (color) and Darcy velocity (arrows) in the model for the upper crust (Fig. 3.17): Doubling of the length of an arrow corresponds to a tenfold increase in velocity. The permeability of the mid-crust is 10^{-18} m^2 (see Table 3.5 for further parameter values).

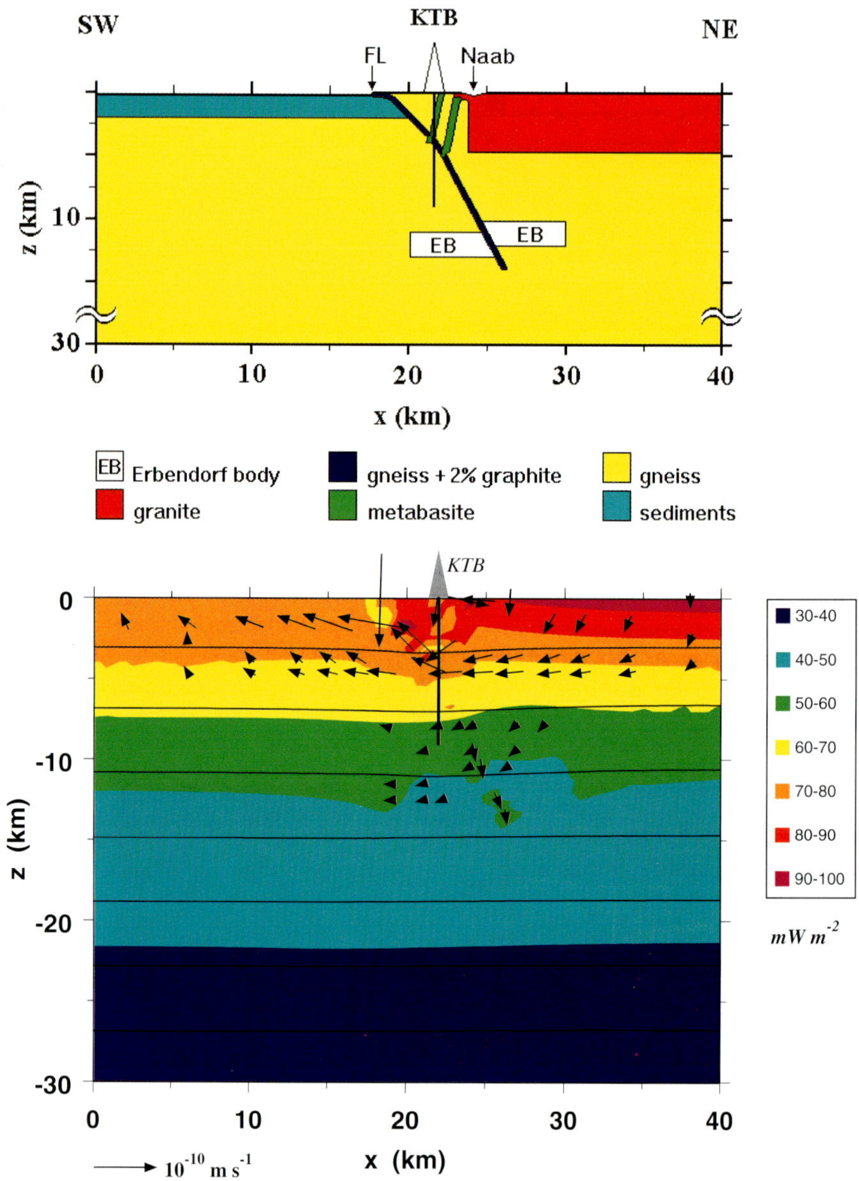

Plate 17. Top: Vertical, 2-D, conceptional model for simulating regional heat transport in the entire crust – this model is similar to that shown in Fig. 3.17 but extends to 30 km. Bottom: Temperature (isolines in °C), heat flow density (color) and Darcy velocities (arrows): Doubling of the length of an arrow corresponds to a tenfold increase in velocity. The permeability of the mid-crust is 10^{-18} m^2, the heat production rate of the lower crust 1 μW m^{-3} (see Table 3.5 for further parameter values).

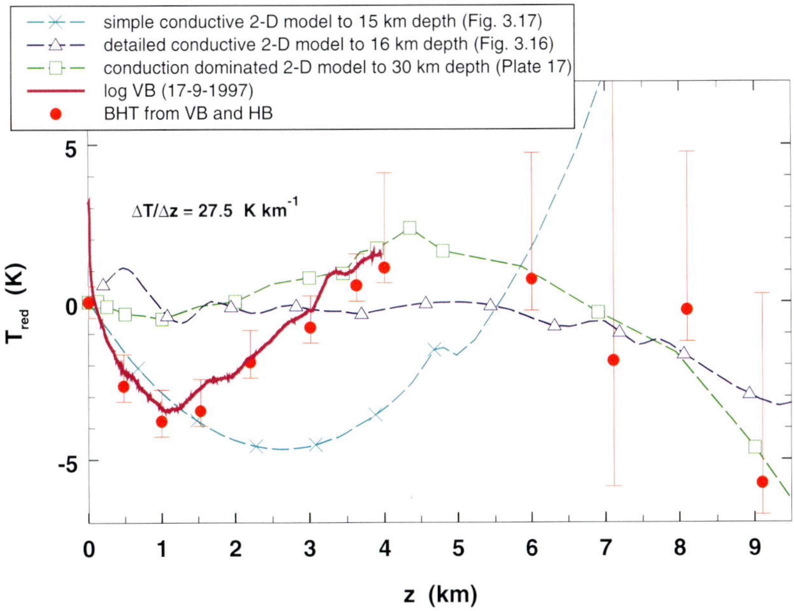

Plate 18. Comparison of reduced temperatures in the KTB boreholes with various model results. The plotted temperature values are reduced by a constant vertical temperature gradient $\Delta T/\Delta z$=27.5 K km^{-1} and mean gound temperature values T_0 equal to 7.5, 8, and 7.4 °C chosen for the log and BHT data, the detailed conductive model (triangles), and the simple conductive as well as the conduction dominated models (crosses and boxes), repectively.

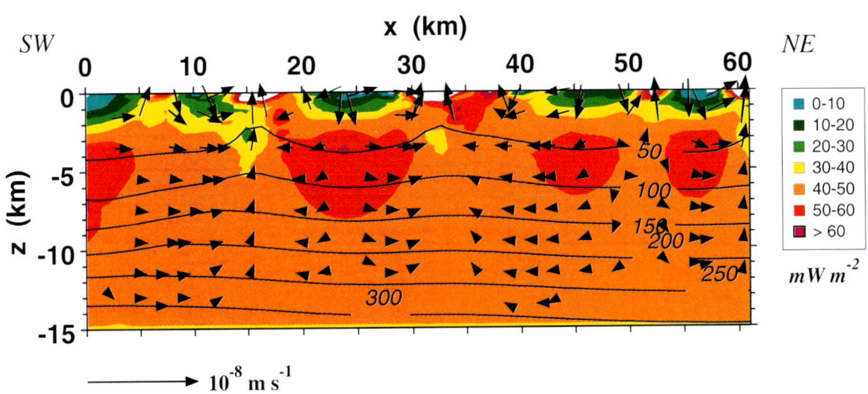

Plate 19. Temperature (isolines in °C), heat flow density (color) and Darcy velocity (arrows) in model 'SEVENTH_40' (Table 3.8): Doubling of the length of an arrow corresponds to a tenfold increase in velocity. See Tables 3.6 and 3.7 and Figs. 3.21 and 3.22 for the parameter values.

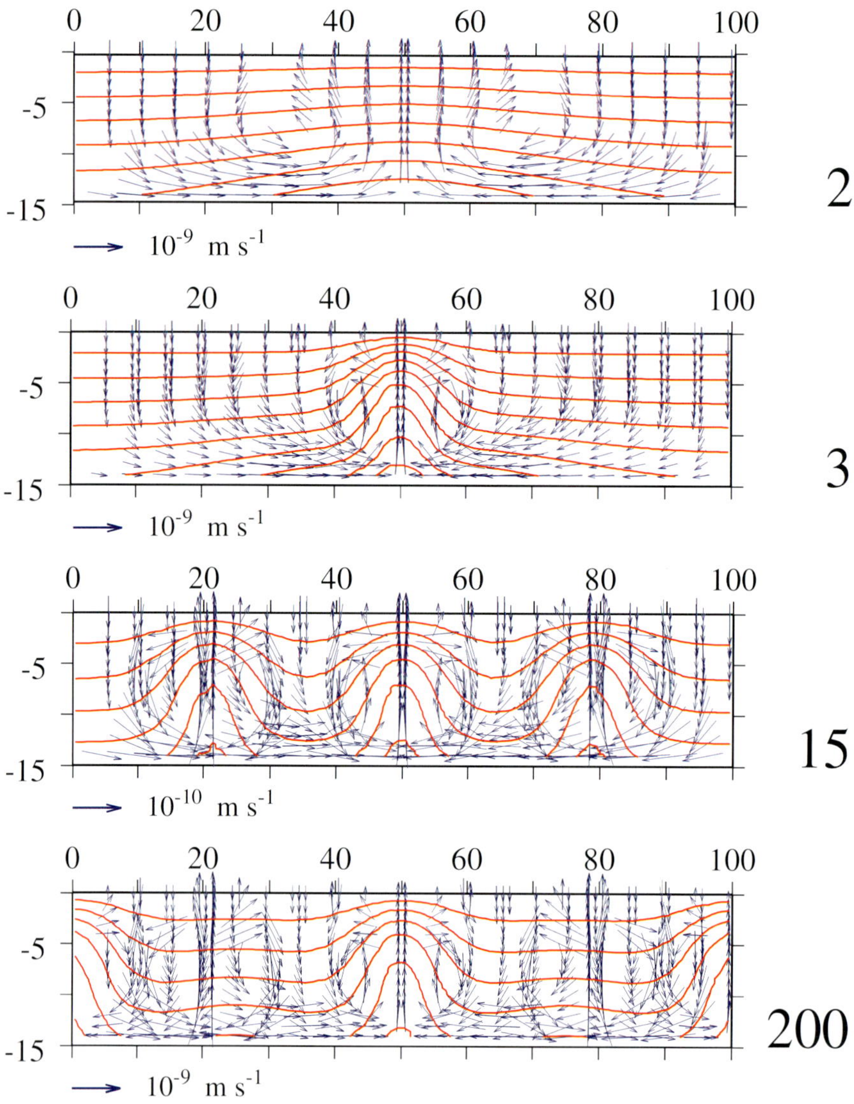

Plate 20. Temperature and Darcy velocity in the homogeneous model I (Fig. 4.5; $k = 10^{-16}$ m^2). The lower boundary condition for heat flow density at the base of the model is defined by function 2 (Figs. 4.3 and 4.4). Temperature isoline spacing is 50 K beginning at 50 °C. Doubling of the length of an arrow corresponds to a tenfold increase in velocity. Distance (x-axis) and depth (y-axis) are in km, the simulation period is given in Ma on the right.

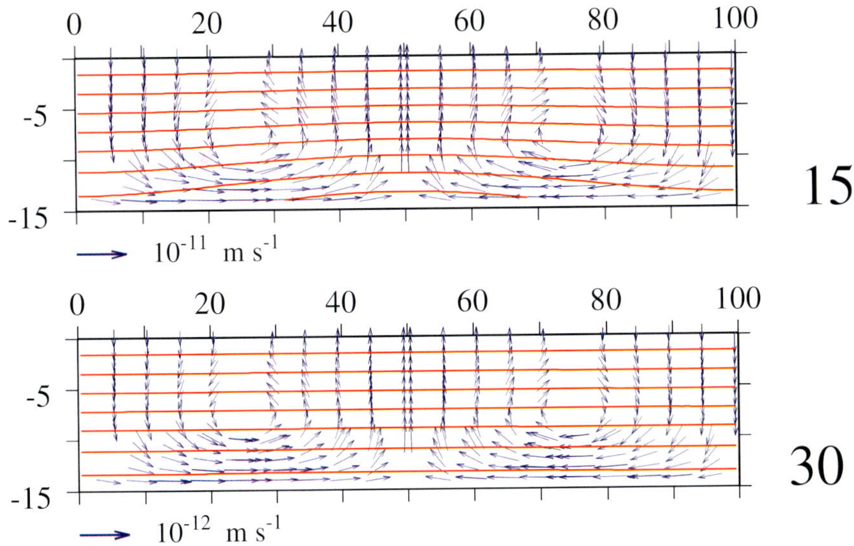

Plate 21. Temperature and Darcy velocity in the homogeneous model I (Fig. 4.5; $k=10^{-17}$ m^2). The lower boundary condition for heat flow density is defined by function 2 (Figs. 4.3 and 4.4). Temperature isoline spacing is 50 K beginning at 50 °C. Doubling of the length of an arrow corresponds to a tenfold increase in velocity. Distance (x-axis) and depth (y-axis) are in km, the simulation period is given in Ma at the right.

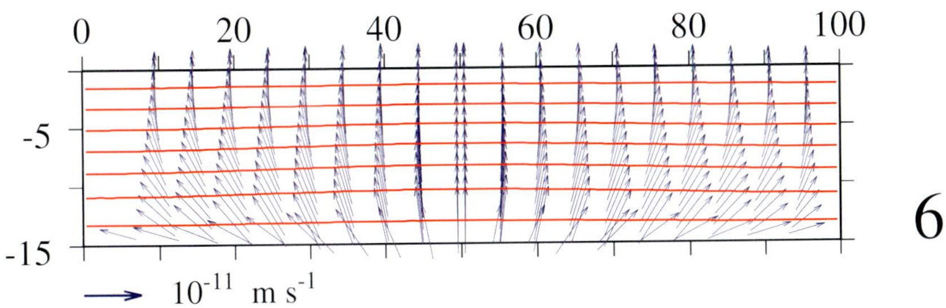

Plate 22. Temperature and Darcy velocity in the homogeneous model I (Fig. 4.5; $k = 10^{-18}$ m^2). In this model, flow is only due to the liberation of 6 wt.% of water of crystallization from the magma shown in Fig. 4.2. The lower boundary condition for mass flow density at the base of the model is defined by function 1 (Figs. 4.3 and 4.4). Temperature isoline spacing is 50 K beginning at 50 °C. Doubling of the length of an arrow corresponds to a tenfold increase in velocity. Distance (x-axis) and depth (y-axis) are in km, the simulation period is given in Ma on the right.

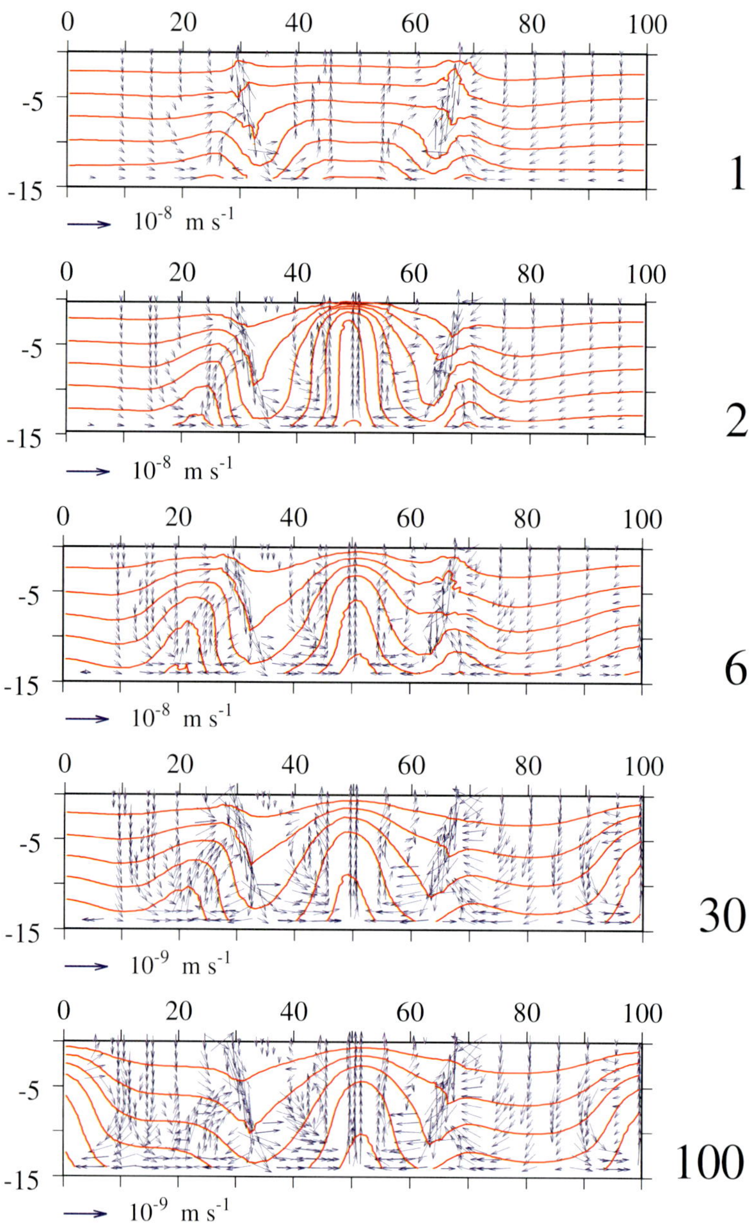

Plate 23. Temperature and Darcy velocity in the structured model II (Fig. 4.5; $k = 10^{-16}\ m^2$). The lower boundary condition for heat flow density is defined by function 1 (see Figs. 4.3 and 4.4). Temperature isoline spacing is 50 K beginning at 50 °C. Doubling of the length of an arrow corresponds to a tenfold increase in velocity. Distance (x-axis) and depth (y-axis) are in km, the simulation period is given in Ma at the right.

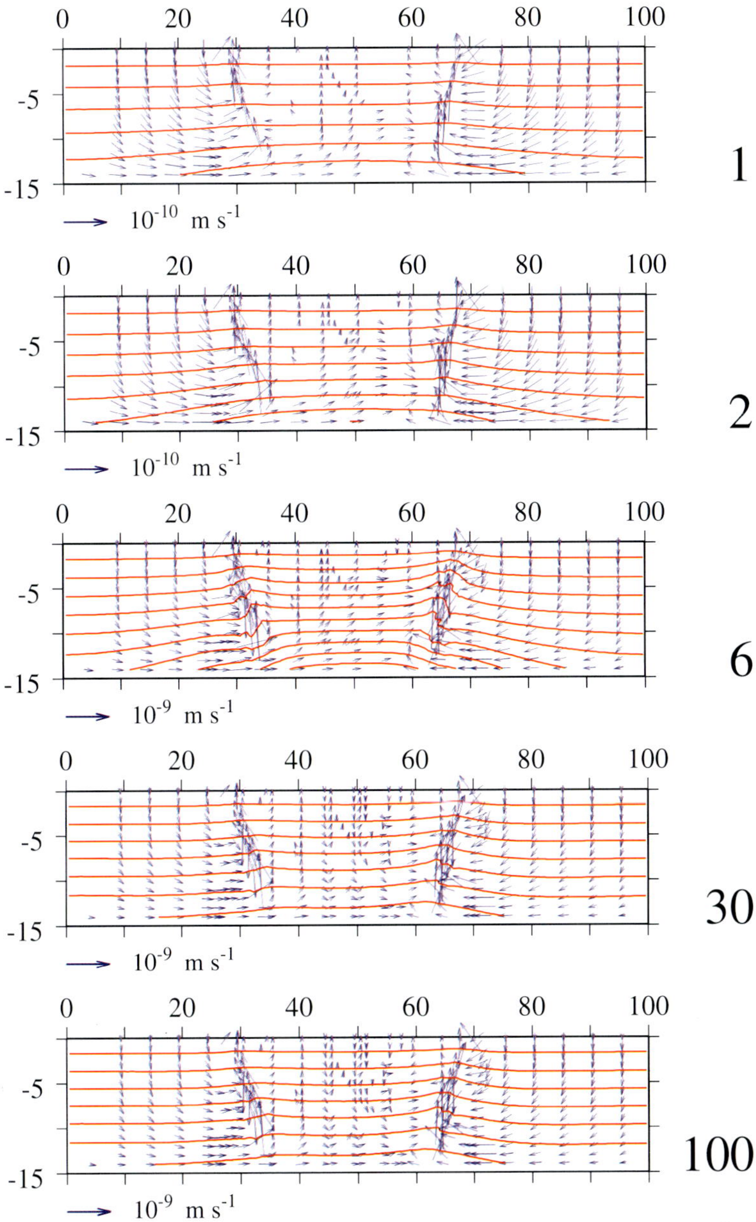

Plate 24. Temperature and Darcy velocity in the structured model II (Fig. 4.5; $k = 10^{-17}$ m^2). The lower boundary condition for heat flow density at the base of the model is defined by function 1 (see Figs. 4.3 and 4.4). Temperature isoline spacing is 50 K beginning at 50 °C. Doubling of the length of an arrow corresponds to a tenfold increase in velocity. Distance (x-axis) and depth (y-axis) are in km, the simulation period is given in Ma on the right.

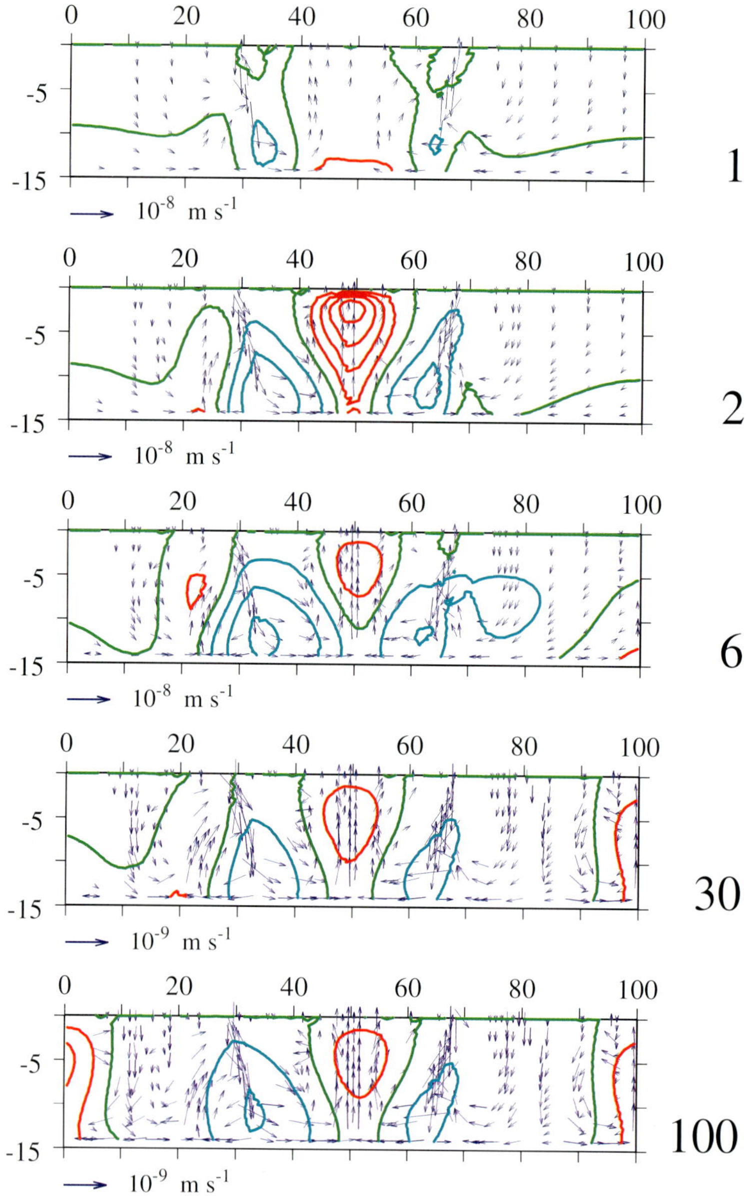

Plate 25. Differences in temperature and Darcy velocity between a simulation with temperature- and pressure-dependent parameters and one with constant coefficients in the structured model II (Fig. 4.5; $k = 10^{-16}$ m^2). The lower boundary condition for heat flow density is defined by function 1 (Figs. 4.3 and 4.4). Temperature isoline spacing is 50 K, the 0 K isoline is green, negative temperatures are blue, positive values are red. Doubling of the length of an arrow corresponds to a tenfold increase in velocity. Distance (x-axis) and depth (y-axis) are in km, the simulation period is given in Ma at the right.

temperature history for central France from two temperature logs recorded in the Limousin region, about 750 km from Bavaria. These boreholes are also deeper than those in Bavaria and Bohemia and permit the reconstruction of a detailed ground temperature history for the last 1000 years. In contrast to Bavaria, this reconstruction shows a warming in central France beginning 200 years ago and lasting until about 50 years ago, followed by a cooling period and a very recent warming. Studies in eastern Canada (e.g., Beltrami and Mareschal 1992) yielded similar trends. The meteorological records appear to correlate for distances of less than 500 km (Hansen and Lebedeff 1987). This is significantly less than the distance between the French and the Bavarian and Bohemian boreholes. Schönwiese et al. (1994) find differences in the climate trends between central and western Europe for the last 100 years.

Thus, the differences in the results of the reconstruction of ground temperature history from boreholes in central and western Europe appear to be consistent with the present understanding of climatic variability. For conclusive evidence for the usefulness of the geothermal reconstruction of ground temperature histories, further studies using a larger data base are necessary. The relationship between air and ground temperatures must also be investigated further and more systematically in order to better interpret the inversion results.

2.2.1.2
The End of the Last Ice Age in the Temperature Log from the KTB Pilot Borehole

A T log was recorded down to a depth of 4 km in the KTB pilot borehole on 17 September 1997, about 8 ½ years after the end of the drilling. The time since any hydraulic experiments were conducted was also large. Thus, the borehole is very likely to have been very close to thermal equilibrium with its surroundings at that time. Therefore, and owing to its depth, this temperature log fulfills very well the conditions required for a reconstruction of climatic events of the distant past. In contrast to the previous example, this is a single temperature log, and hence individual deviations from steady-state, vertical and conductive conditions that are not caused by climatic conditions can have an immediate effect on the results. In order to better suppress any undesired fluctuations in the reconstructed ground temperature history, the attenuated inversion of equation (18) was used. The differences from the unattenuated cutoff version are not very large, however, and are expressed mainly in the form of a somewhat smoother error envelope. The results for various values of the attenuation factor ε show that again a trade-off is required between too much attenuation suffering from too little resolution on the one hand, and higher resolution plagued with oscillations on the other hand. The following figures illustrate this for attenuation factors of $\varepsilon = 0.2$ and 0.4.

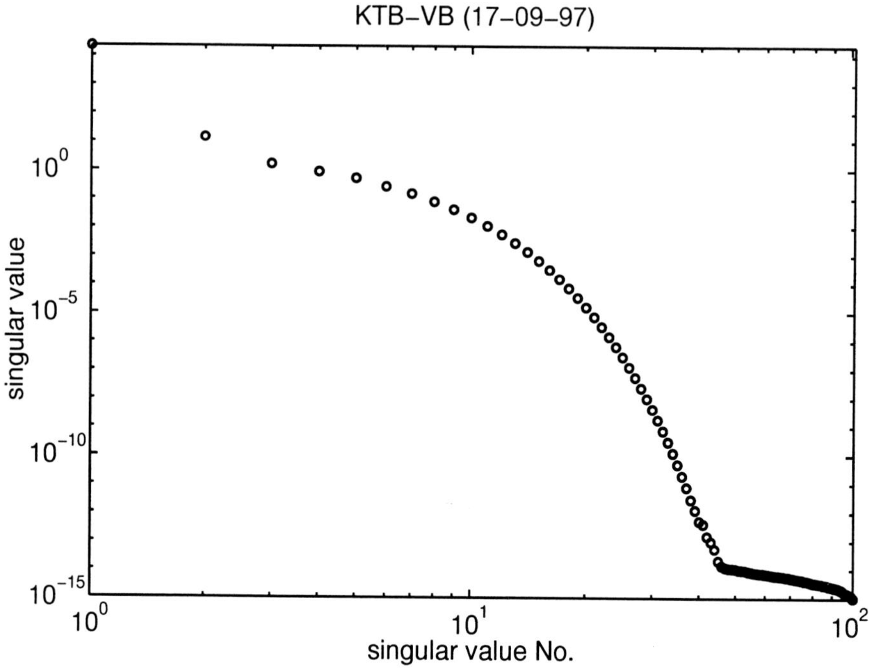

Fig. 2.10. Magnitudes of the 102 singular values resulting from inversion of the temperature log recorded in KTB pilot borehole on 17 September 1997. See text for more details.

The T log was re-sampled at 1 m interval and inverted for the period of 1,000,000 – 100 years before 2000 A.D. using 100 logarithmically equidistant time steps, a thermal conductivity of 2.92 W m^{-1} K^{-1}, a thermal diffusivity of 10^{-6} m^2 s^{-1}, and a radiogenic heat production rate of 1.1 μW m^{-3}. Figure 2.10 shows how quickly the eigenvalues approach zero: Only seven or six out of 102 singular values are larger than 0.2 or 0.4, respectively, and the following values approach zero very rapidly. Thus, it can be seen how unphysical oscillations in the inverted ground temperature history, which are caused by small singular values, will be suppressed by either applying a suitable cutoff value or by including an attenuation factor $\varepsilon > 0$ in equation (18).

The fit of the model obtained with this inversion using an attenuation factor of $\varepsilon = 0.2$ is shown in Plate 8, together with the perturbations in temperature and heat flow density, which are related to the ground temperature history. Considering the magnitude of the temperature signal, the model misfit on the order of 10 % for a large portion of the log seems to be acceptable. This is obtained, however, at the expense of fluctuations in the first 1000 years of the reconstructed ground temperature history (not shown in Plate 8), which, of course, have nothing to do with natural climate fluctuations. The corresponding reconstructed ground temperature history is plotted in Plate 9 with paleo-climatological proxy data for Germany and Switzerland (Zoth and Hänel 1988). Also shown is the model response to a cubic spline interpolation of this proxy data to the times of the ground temperature history. It is calculated according to equation (20) as the product of the model resolution matrix and the vector of this interpolated proxy temperature history.

The strongest signal in the inverted ground temperature history is a post-glacial, 10 K warming starting at about 40,000 before the year 2000, that reaches its maximum at about 1000 B.C. This signal is associated with the largest singular value besides the ones that characterize the average surface temperature and the reference temperature gradient. This can be seen by subsequently increasing the attenuation factor ε until the inverted ground temperature history becomes constant with time. This is because the Pleistocene temperature perturbations recorded in the 4 km deep KTB pilot borehole reach their maximum of about 2.7 K at a depth of about 1500 m and still amount to about 1.3 K at bottom hole (Plate 8). There is acceptable agreement between inverted ground temperature history and the proxy data, both in respect to the amplitude and the time of the onset of the post-glacial warming. The inverted ground temperature history compares well with the model response to the proxy data. Owing to diffusion, both follow the trend of the proxy data in an smoothed and phase-shifted sense. Thus, peaks of the proxy time series cannot be recovered.

The corresponding results for a larger attenuation factor of $\varepsilon = 0.4$ are shown in Plates 10 and 11. As expected, the quality of the fit to the data decreases (Plate 10). The error is generally larger than for the smaller attenuation factor of $\varepsilon = 0.2$, frequently exceeding a value of 10 % of the temperature signal. This goes along with a stronger attenuation of the fluctuations in the first 1000 years of the ground temperature history (not shown). However, at the same time the maximum amplitude of the post-glacial warming decreases to about 8 K (Plate 11), which appears a bit low.

These two examples clearly illustrate the difficulties encountered when the attempt is made to solve this complicated underdetermined and overdetermined inverse problem. Extraction of the information about temperature variations at the Earth's surface, which is undoubtedly present in suitable temperature logs, cannot be automated. Additional information is required, as well as the personal judgment of the researcher.

3 Identification of Diffusion and Advection: Two Case Studies from the Deepest Basement Drill Holes, the Kola and KTB Deep Boreholes

The desired result of most geothermal borehole logs (T logs) is an, at least piecewise, linear temperature profile. In a horizontally layered part of the earth's crust, such a profile normally corresponds to purely conductive, steady-state conditions. Such conditions are required for a number of standard interpretation methods for geothermal studies, for example, one-dimensional extrapolation of temperature towards depth, determination of heat flow density from a "Bullard Plot", i.e. calculation of a linear regression curve of temperature as a function of the thermal resistance (Bullard 1939), or determination of the relationship between heat flow density and the heat production rate, which is used to estimate (a) the thickness of the Earth's crust in which radiogenic heat production occurs and (b) the corresponding mantle heat flow density (Birch et al. 1968, Lachenbruch 1968, Roy et al. 1968). These methods lead to an incorrect interpretation when the distribution of thermal conductivity or heat production rate is relatively complex, as well as when transient conditions prevail or other heat transport processes occur at the same time. The probability for this is increased by a trend to the use of shallow boreholes for geothermal studies (Chapman et al. 1984). The use of mostly shallow boreholes favors the occurrence of advective disturbances and makes it more difficult to recognize the presence of other heat transport processes.

For this reason, both the deep drilling projects in the Soviet Union at end of the 1980s (Kremenetsky and Krivtsov 1991), especially the SG-3 borehole on the Kola Peninsula (at 12,261 km total depth, it is still the deepest borehole in the world), and the German Continental Deep-Drilling Program in Bavaria (Emmermann and Wohlenberg 1989) met with considerable interest. These boreholes opened, for the first time, the fascinating possibility to test the interpretation methods mentioned above using a profile down to great depth in the crystalline basement, as well as to study in situ the different heat transport processes at depths accessible only indirectly until then. One of the main results is a variation in heat flow density with depth observed in both boreholes in a form previously unexpected. Despite the progress made, these variations are still not yet completely understood. The present results are presented in the next two sections. Differences between Kola and KTB are due less to the methods than the availability of good and representative data.

3.1 KTB: Study of Heat Transport Under Steady-State Conditions

Temperature logs and bottom hole temperature (BHT) measurements in the KTB pilot and main boreholes show a rather uniform temperature increase of 27.5 ± 1 K km^{-1} below a depth of about 2 km. Equilibrium temperature is 118.6 °C in the KTB pilot borehole (cf. Plate 2) and will be around 260 °C at 9.1 km in the KTB main borehole (cf. Plate 3) (Clauser et al. 1997).

A considerable problem for understanding heat transport processes in the area of the KTB borehole is the observation in the pilot borehole at a depth of 500 m of an increase in the vertical heat flow density of more than 50 %: from about 52 mW m^{-2} above 500 m to a mean of 83 mW m^{-2} below that depth. This observation caused serious problems for the temperature extrapolation for the pilot borehole, since the geothermal measurements preceding the drilling of the pilot borehole were all made in shallow boreholes with depths of less than 500 m (Burkhardt et al. 1989). There are three mechanisms that can explain this unexpected increase in the mean vertical heat flow density: (1) refraction of the vertical heat flow by a large lateral difference in thermal conductivity, (2) reduction of the near-surface heat flow density by recharge of meteoric water in a regional convection system controlled by topography and (3) lowering of the near-surface temperature gradient by downwards diffusion of climatic variations of the mean surface temperature. The last mechanism was discussed in chapter 2. The first two mechanisms are discussed in section 3.1.1.

The data from the main borehole raise new questions: A one-dimensional extrapolation of the heat flow density at 8.4 km yields values between 43 and 68 mW m^{-2} for a depth of 30 km, assuming a constant radiogenic heat production rate of 1 μW m^{-3}. This corresponds to temperatures higher than 800 °C, even when the uncertainties in the mean thermal conductivity of the crust assumed for the extrapolation are taken into consideration. The solidus of most rocks, especially of acid to intermediate rocks, is in this temperature range when water is present (e.g., Thompson 1992). This temperature gradient would lead to liquidus temperatures at depths of 45 – 55 km, which would cause melting of the rocks. Seismic studies yield no indication of the presence of partial melts in the lower crust. Hence, such high temperatures are not very probable for these crustal depths. For this reason, a decrease in the vertical temperature gradient is expected. This could be caused, for instance, by an elevated radiogenic heat production rate or other heat sources below the bottom of the borehole. Such sources, however, can only be due to transient processes, for example, sub-recent volcanic or intrusive activity or the exothermic release of binding energy accompanying the hydration of acid crustal rocks. Neither process has occurred in Bavaria in recent times, however. When the geological age of these rocks is taken into consideration, it does not appear very probable that such processes can account for the observed differences in heat flow density, even if they occurred in the geological past (Clauser et al. 1997). Convection of fluids throughout the crust or the focusing of the regional conductive heat flow by lateral heterogeneities in thermal conductivity could also be responsible for the relatively high temperature gradients observed. This problem cannot be completely clarified at the present time, in spite of the comparatively large data base. Numerical modeling provides, however, some indication of the permeability of the crust, heat flow densities in the lower crust, and the thermal conductivity distribution required for the various above-mentioned processes. This is discussed in section 3.1.2.

3.1.1
Does Groundwater Recharge Drive a Near-Surface Circulation System in the Study Area?

3.1.1.1
The Regional Temperature Field

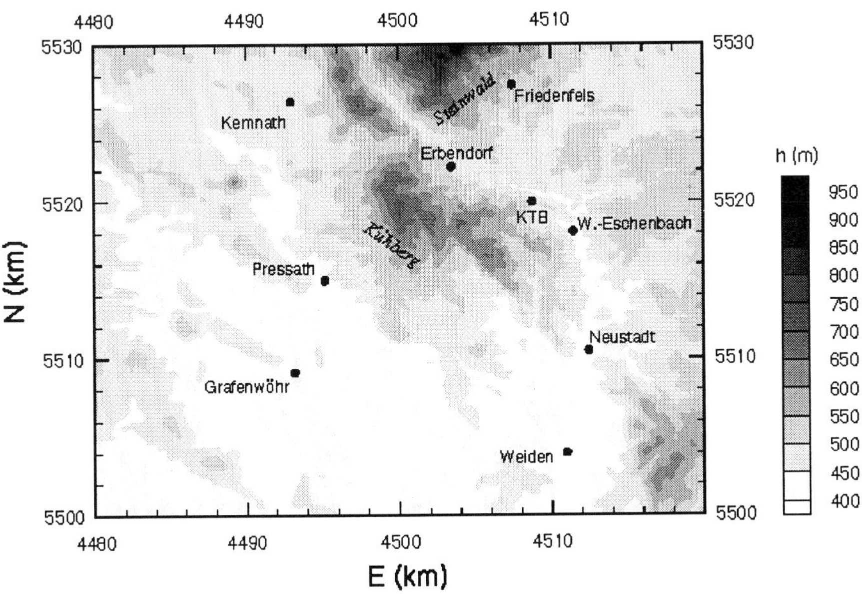

Fig. 3.1. Topography in the area around the KTB borehole in Bavaria (E, N: easting and northing in the Gauss-Krüger system; h: elevation in m above m.s.l.); data from digital terrain model VM 2255B-3991 by permission of the Bavarian Topographical Survey, Munich.

If the increase in vertical heat flow density at 500 m in the pilot borehole is not a purely local phenomenon but is caused by a regional groundwater circulation system, then this increase must be recognizable in the temperature logs of other boreholes in the region. All in all, 25 boreholes are available for comparison (Table 3.1). Twenty-one of them are in the crystalline basement of the Bohemian Massif east of the Frankonian Line and four are in Permian and Mesozoic sediments to the west of that line (Jobmann and Clauser 1994). Unfortunately, this data density is too low for an area of about 60 × 80 km to plot temperature maps for selected depth levels. A slight regional trend, however, can be recognized from the correlation of temperatures in these boreholes at a certain depth. Therefore, Figure 3.3 shows the temperature in these boreholes at a depth of 100 m below ground surface.

Table 3.1. Number, name, elevation h, easting, northing, maximum depth z and temperature T at a depth of 100 m in the 25 boreholes in Fig. 3.3 (italics: boreholes used for Péclet number analyses – see Table 3.2).

No.	borehole	h [m]	easting [km]	northing [km]	z [m]	$T_{100\ m}$ [°C]
1	Weißenstein	665	4477.97	5554.87	231	7.9
2	Alexanderbad	600	4501.11	5541.61	110	8.2
3	*Egerer Waldhaus*	*703*	*4530.70*	*5533.36*	*200*	*8.4*
4	*Wülfersreuth*	*720*	*4482.84*	*5546.84*	*200*	*8.4*
5	*Griesbach*	*720*	*4535.90*	*5525.45*	*301*	*8.5*
6	Poppenreuth	642	4532.70	5531.12	239	8.6
7	Egerer Waldhaus	700	4530.10	5533.82	200	8.6
8	Herrmannsreuth	745	4533.24	5521.53	11	8.8
9	*Neusorg*	*636*	*4499.75*	*5535.60*	*301*	*8.9*
10	Kondrau	486	4521.11	5540.14	235	9.0
11	*Weissenstadt*	*635*	*4493.33*	*5550.14*	*220*	*9.0*
12	*Röthenbach*	*536*	*4510.70*	*5546.44*	*140*	*9.1*
13	*Püllersreuth*	*548*	*4509.68*	*5516.32*	*301*	*9.2*
14	Remmelberg	570	4519.40	5503.86	140	9.4
15	Friedersreuth 10/90	530	4502.28	5515.70	250	9.9
16	*Falkenberg NB3,*	*507*	*4514.42*	*5524.71*	*301, 500*	*9.9*
17	Sybillenbad H1	542	4531.55	5537.31	315	10.0
18	Erbendorf 5/89	590	4501.02	5523.44	130	10.1
19	Sybillenbad A	542	4531.55	5537.31	250	10.3
20	*KTB pilot borehole*	*514*	*4508.59*	*5519.86*	*4001*	*10.5*
21	Lengenfeld II	549	4520.00	5521.14	417	10.6
22	Dobrigau 7/89	507	4520.85	5532.78	171	10.9
23	Weiden T1	397	4511.74	5505.54	1222	11.7
24	*Bayreuth SW*	*351*	*4473.12*	*5535.88*	*1122*	*12.7*
25	Obernsees 1	381	4455.53	5531.03	685	17.0

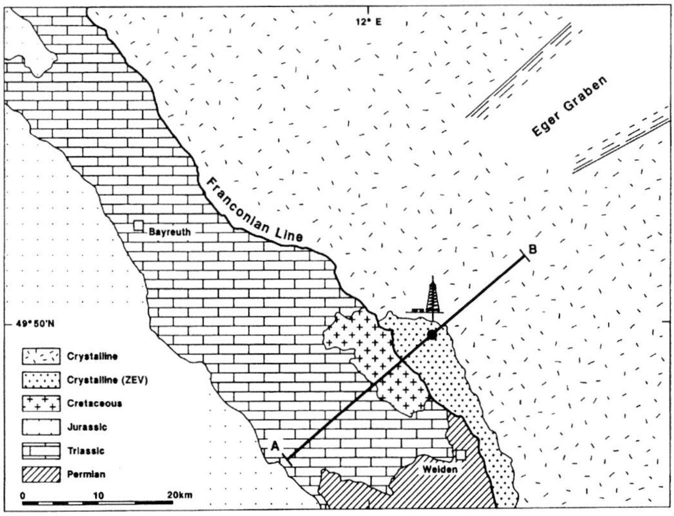

Fig. 3.2. Simplified geological map of the area around the KTB borehole.

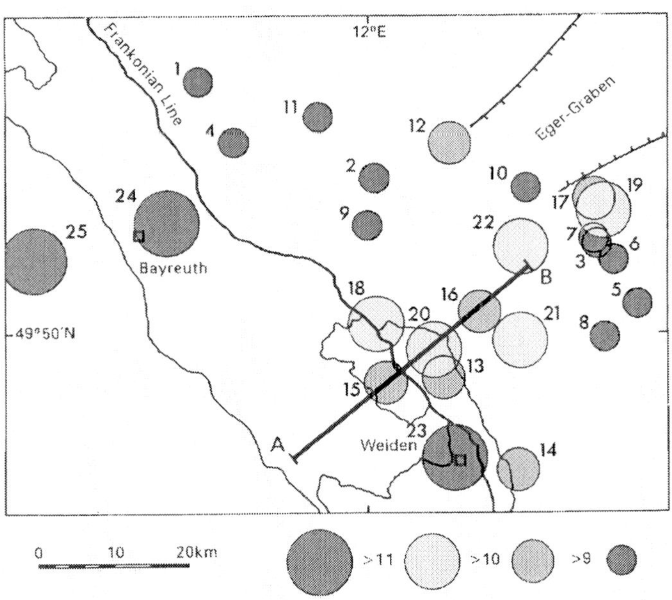

Fig. 3.3. Temperatures in °C at a depth of 100 m below ground surface recorded in 25 shallow boreholes in the area around the KTB borehole (see Table 3.1): The boundaries of the stratigraphic units in Fig. 3.2 are shown for orientation.

The Frankonian Line (FL) also correlates approximately with the increase in mean elevation of 125 m between the area of Mesozoic sediments to the southwest of the FL and the Bohemian Massif (Fig. 3.1). The only exception is the area around the Friedersreuth borehole in the Cretaceous (No. 15 in Table 3.1 and Fig. 3.3), which is higher than the other parts of the area of Mesozoic sediments. The temperatures of 11.7 – 17.0 °C in three boreholes in the Mesozoic sediments (Nos. 23 – 25 in Table 3.1 and Fig. 3.3) are distinctly above the 7.9 – 10.5 °C of the other 22 boreholes east of the Frankonian Line. Over a distance of 15 km across the Frankonian Line, this temperature difference is between 1.2 K and 4.3 K for any one borehole. This cannot be explained only by the variation of air temperature with elevation: For a difference in elevation of 200 m, the adiabatic temperature gradient explains temperature differences of only 1 – 2 K for dry or water-saturated air (Jobmann and Clauser 1994). This is distinctly less than the observed temperature differences. This could be evidence for lateral heat transport, possibly driven by a regional system of groundwater circulation from the Bohemian Massif across the Frankonian Line into the area of Mesozoic sediments.

3.1.1.2
Vertical Péclet Number Analysis

A regional groundwater circulation system would affect both temperature field and heat transport, and hence heat flow density. There would be groundwater recharge in the Bohemian Massif region, and discharge in the lower sedimentary foreland. Initial evidence for this is provided by the observation that the three boreholes in the Mesozoic sediments – at Obernsees, Bayreuth and Weiden (Table 3.1) – are all artesian. Regional recharge would lead to a systematic lowering of the heat flow density observed in the boreholes in the Bohemian Massif at the depth of this circulation system and to an increase below that depth. This is exactly what is found in the KTB pilot borehole. Therefore, it should be checked whether evidence for this can also be found in the data from the shallow boreholes.

The simplest model for testing this hypothesis is advection of heat by constant, vertical groundwater flow in a homogeneous, isotropic medium. This hypothesis can be tested by subjecting the temperature logs to a vertical Péclet number analysis (Pe analysis, equation 4). In a Pe analysis, depth intervals with a thickness $L = z_B - z_0$ and a temperature difference $\Delta T = T(z_B) - T(z)$ are analyzed from z_0 to z_B (Fig. 3.4). An expression for the variation of temperature with depth is obtained by solving the heat transport equation with the corresponding boundary conditions for the simple 1-D model described above. By plotting normalized temperature versus normalized depth, the value of Pe can be determined by curve fitting (Bredehoeft and Papadopulos 1965).

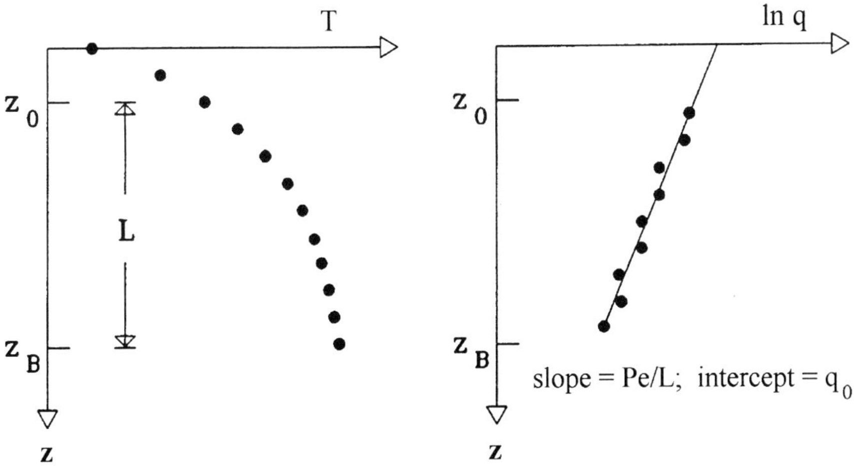

Fig. 3.4. Vertical Péclet number analysis: Vertical Darcy flow within the depth interval $L = z_B - z_0$ gives rise to advection of heat. This can be seen both in the nonlinear T log and in the linear trend of ln q(z).

Alternatively, the ratio of Pe to L can be determined from a linear regression of the curve of temperature gradient versus temperature (Mansure and Reiter 1979). Equation (22) defines the corresponding expressions for heat flow density q(z) as a function of depth z (with the positive axis downwards). These equations can be used to determine Pe/L and $q_0 = q(z_0)$, either from an exponential regression curve for heat flow density and depth (Sass and Lachenbruch 1982) or from a linear regression curve for the logarithm of heat flow density and depth (Clauser and Villinger 1990) (Fig. 3.4). This alternative approach also permits horizontal layers of differing thermal conductivity to be taken into consideration. However, owing to the complicated structure of the crystalline basement of the Bohemian Massif, this was not attempted when analyzing these temperature logs.

$$q(z) = q(z_o)e^{-(z-z_o)\frac{Pe}{L}} \quad \text{or} \tag{22}$$

$$\ln(q(z)) = \ln(q(z_o)) - (z - z_o)Pe/L.$$

On the other hand, a theoretical T log can be calculated for a homogeneous, isotropic half-space, that corresponds to the measured temperature $T_0 = T(z_0)$, as well as to the parameters q_0 and Pe/L determined using equation (21) (Mansure and Reiter 1979):

$$T(z)=\frac{q_o/\lambda}{Pe/L}e^{-(z-z_o)\frac{Pe}{L}-1}+T_o e^{-(z-z_o)\frac{Pe}{L}}. \tag{23}$$

This permits not only the quality of the regression to be checked, but also the agreement between measured T log and model prediction. Additional information is required, however, to determine the Péclet number of the system and the conductive heat flow density at the base, $q_0 = q(z_0+L)$, uninfluenced by vertical advection. This is because the regression yields only the ratio Pe/L, but not the individual parameters. Consequently, L must be estimated from ancillary information.

The abrupt increase in heat flow density in the KTB pilot borehole indicates that 500 m is the depth to which a shallow recharge system has any influence. This is supported by the temperature logs from the KTB pilot borehole, which indicate an elevated fracture porosity in this area (Jobmann and Clauser 1994). Möller et al (1997) also report a transition from mobile to nearly immobile formation water at a depth of around 500 m, determined on the basis of isotopic analysis of borehole fluids. For the temperature logs from boreholes in the crystalline basement of the Bohemian Massif, L = 500 m is therefore assumed to be the maximum thickness of the recharge regimes. For the Bayreuth borehole, which is the only one of the three in the Mesozoic sediments that is suitable for a Pe analysis, two limiting values were used to determine L: L = 700 m for the length of the curved interval in the T log above the Bunter aquifer between 850 and 1000 m depth, and L = 1300 m for the depth to the crystalline basement (Gudden and Schmidt 1985; Schmoll et al. 1989).

Analysis of the Péclet numbers for the top 500 m of the temperature log measured in the KTB pilot borehole shows very little difference (a maximum of 130 mK) between the model prediction and the measured data. But when the T log down to 2200 m is used for the analysis, the agreement is considerably poorer, and the temperature differences are as large as 2 K. Analysis of the log from 500 to 2200 m still yields an unsatisfactory difference of as much as 1 K. This is a further indication that a thermally relevant flow system, if it exists, is confined to the top 500 m. The Péclet number obtained from this analysis is Pe=-0.35 (Table 3.2). According to equations (4) and (5), this means that an additional advective heat transport component exists that amounts to 1/3 of the conductive component, or that advection accounts for more than ¼ of the total heat transport. The corresponding Darcy velocity v, calculated using equation (4) is -13.0 mm a^{-1}. The tracer velocity u can be calculated from the Darcy velocity and the porosity Φ as follows: $u = v/\Phi$. Thus, for the porosities of 1 – 2 % in this area (Berckhemer et al. 1997), meteoric water requires 385 – 770 years to travel from the Earth's surface to a depth of 500 m in the crystalline basement.

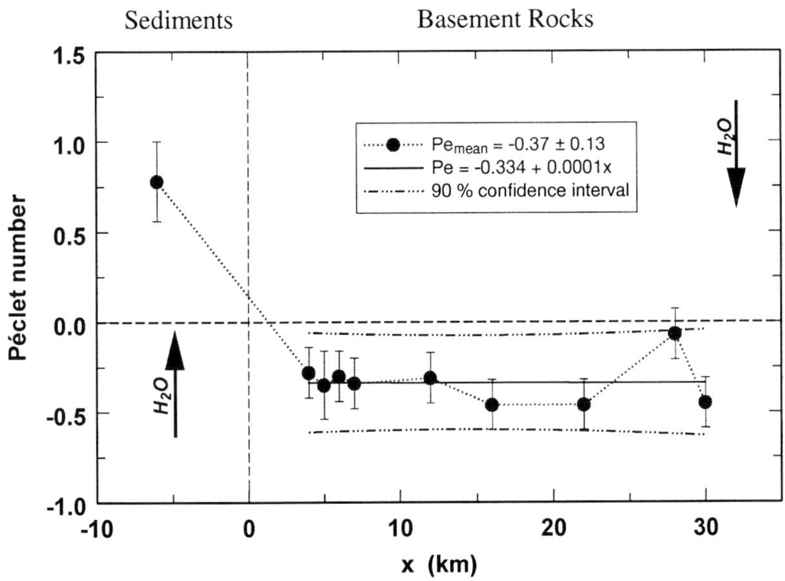

Fig. 3.5. Péclet numbers from Pe analyses projected onto profile A—B perpendicular to the Frankonian Line (Fig. 3.2): From left to right, the boreholes are ordered as in Table 3.2 .

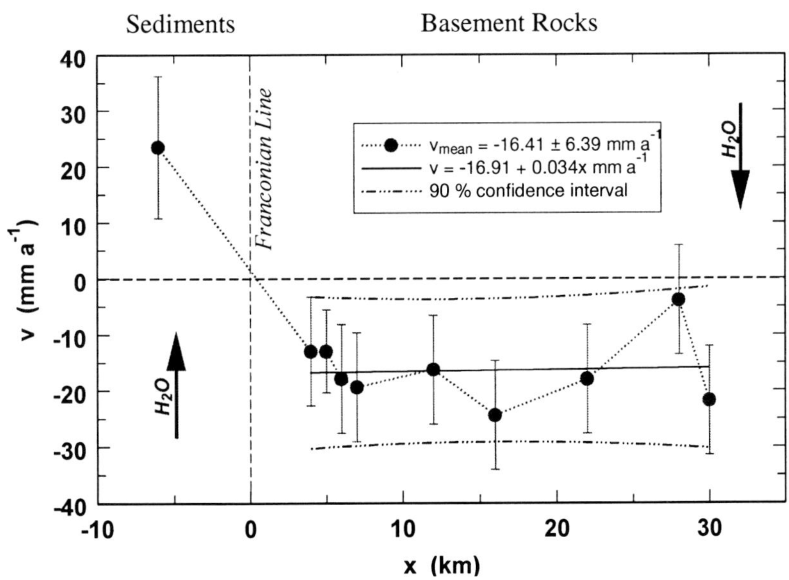

Fig. 3.6. Darcy velocity v from Pe analyses projected onto profile A—B perpendicular to the Frankonian Line (Fig. 3.2): From left to right, the boreholes are ordered as in Table 3.2.

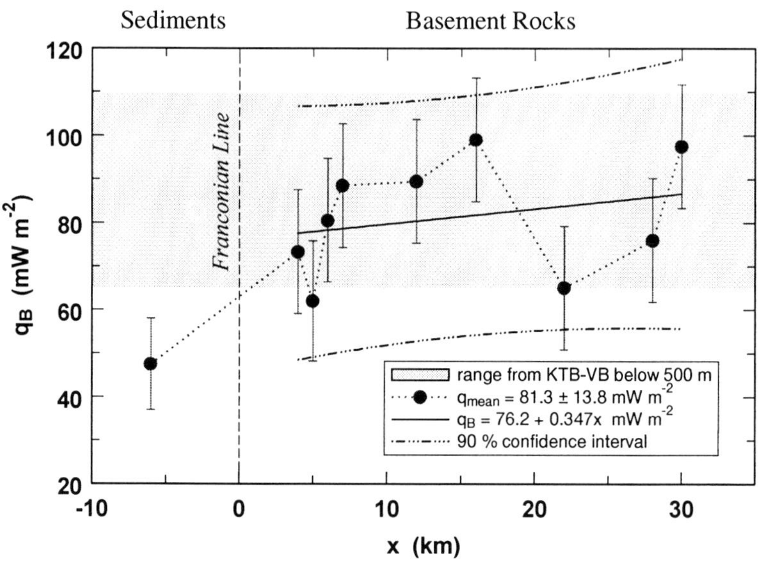

Fig. 3.7. Heat flow density q_B at the base of the model from Pe analyses projected on the profile A—B perpendicular to the Frankonian Line (Fig. 3.2): From left to right, the boreholes are ordered as in Table 3.2. Shaded area indicates the variation of heat flow density in the KTB pilot borehole in the depth interval from 0.5—4 km.

Table 3.2 Parameter values and results of Péclet number analysis for the ten boreholes studied. Name, number and maximum logged depth z are given in the first three columns. The remaining seven columns give thermal conductivity λ (derived from the values of Burkhardt et al. 1989), Pe/L from the robust linear regression, apparent heat flow density q_0 at the top of the analyzed depth interval, Darcy velocity v = Pe/L · λ/(ρc)$_f$ (with the constant $(\rho c)_f$ = 4.1665 MJ m^{-3} K^{-1}), length L of the depth interval for advection and the Péclet numbers Pe and heat flow density at the base, q_B, calculated from it (see text for further explanations).

borehole	No.	z [m]	λ [W m^{-1} K^{-1}]	Pe/L [km^{-1}]	q_0 [mW m^{-2}]	v [mm a^{-1}]	L [km]	Pe	q_B [mW m^{-2}]
Bayreuth SW	24	1122	2.80	+1.11	103.1	+23.5	0.70	+0.78	47.5
Püllersreuth	13	301	3.08	-0.56	55.8	-13.0	0.5	-0.28	73.3
KTB pilot hole	20	4001	2.46	-0.70	46.7	-13.0	0.5	-0.35	62.0
Wülfersreuth	4	200	3.89	-0.61	60.8	-17.9	0.5	-0.30	80.5
Neusorg	9	301	3.80	0.67	66.2	-19.4	0.5	-0.34	88.5
Falkenberg	16	500	3.43	-0.63	67.5	-16.3	0.5	-0.31	89.5
Weissenstadt	11	220	3.37	-0.97	62.7	-24.4	0.5	-0.46	99.0
Röthenbach	12	140	2.62	-0.91	42.6	-18.0	0.5	-0.46	65.0
Griesbach	5	301	3.44	-0.15	71.0	-3.9	0.5	-0.07	76.0
Poppenreuth	6	239	3.17	-0.91	68.4	-21.8	0.5	-0.45	97.5

Péclet number analyses were made for all suitable shallow boreholes as well as for the KTB pilot borehole. In three cases, the result could not be interpreted: A clear indication that the simplifying assumptions are not fulfilled. However, eleven boreholes yielded reasonable results: ten in the Bohemian Massif and the Bayreuth borehole in the Mesozoic sediments. Two of the first 10 boreholes are close neighbors (Falkenberg NB3 and PB7) and yielded nearly identical results; therefore, they are not listed separately. All of the calculations were made using constant values of $\rho = 997$ kg m^{-3} and $c_p = 4179$ J kg^{-1} K^{-1}. The "robust" variant of the linear regression (see, for example, Press et al. 1992) proved to be better than the standard procedure, because the robust linear regression better reflects the trend of most of the data and is less affected by outliers. The results are compiled in Table 3.2; the vertical Péclet numbers, Pe, the corresponding vertical Darcy velocities, v, and heat flow density at the base, q_B, are plotted in Figures 3.5 – 3.7 along a profile perpendicular to the Frankonian Line (profile A—B in Figs. 3.2 and 3.3). The results for all the boreholes in the area of the Bohemian Massif indicate groundwater recharge from the surface. There is, however, no lateral trend in the values for Pe, v and q_B. All three parameters are adequately represented within the limits of precision by their mean values: $Pe_{mean} = -0.37 \pm 0.13$, $v_{mean} = -16.4 \pm 6.4$ mm a^{-1} and $q_{mean} = 81.3 \pm 13.8$ mW m^{-2}. A strong upwelling of formation water can be recognized in the data from the Bayreuth borehole in the Mesozoic sediments, however. Although this is observed in only one borehole, it is considered to be significant, because all of the other boreholes in the Mesozoic sediments are artesian. The other boreholes, however, could not be used for a Péclet number analysis for various reasons. The results of this Péclet number analysis may be considered, therefore, a further indication of a regional groundwater circulation system. Such a system would lead to a lowering of the surface heat flow density in the Bohemian Massif region. We will now investigate the extent to which this hypothesis is supported by the results of model calculations.

3.1.1.3 Numerical Modeling

Two-dimensional steady-state models are used to study the potential effects of advective heat transport. These calculations are based on models that greatly simplify the actual geological conditions. For this reason they are particularly suitable for illustrating the regional relationships and less suitable for interpreting data that is only locally significant. The computations were made using the SHEMAT finite-differences program to solve the mutually coupled equations for groundwater flow and heat transport in a porous medium (see Appendix 1). Both the rock and the fluid properties were updated according to the prevailing P-T conditions during the computations, in particular density, viscosity and thermal conductivity of the water and the thermal conductivity of the rock. Further physical, mathematical and numerical details about the simulation program are described in Clauser (1988a) and Clauser and Villinger (1990) and are therefore not repeated here.

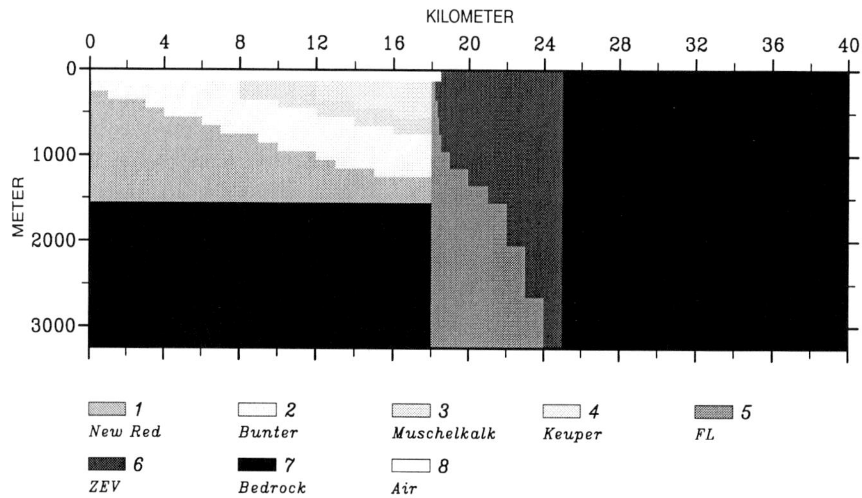

Fig. 3.8. Vertical 2-D conceptual model for simulating near-surface advection systems (Jobmann and Clauser 1994) along profile A—B (Fig. 3.2) perpendicular to the Frankonian line (KTB is at km 22). The geological units are highly simplified after Hirschmann (1993).

Table 3.3. Parameter distribution for the model for shallow depths (Fig. 3.8): thermal conductivity λ, heat production rate H, and porosity Φ.

domain	λ [W m^{-1} K^{-1}]	H [μW m^{-3}]	Φ [%]
1: Rotliegend	2.2	0.40	10
2: Bunter	2.2	0.40	15
3: Muschelkalk	2.2	0.40	15
4: Keuper	2.2	0.40	10
5: Frankonian Line	2.6	0.75	5
6: ZEV area (a high-grade metamorphic complex)	3.0	1.10	1
7: basement	3.5	6.70	3
8: air	0	0	0

The top boundary of the model is set to a constant hydraulic potential corresponding to the elevation and the temperature is kept constant at 7.4 °C. Neither water nor heat may flow across the lateral boundaries of the model. Whereas this is also the case for water at the lower boundary, there is a heat flow density of 80 mW m^{-2} across this boundary. The thermal parameter values for the sedimentary units (1 – 4) and the crystalline basement (5 – 7) are given in Table 3.3. In the model, the permeability is varied systematically from 10^{-12} to 10^{-14} m^2 in the sedimentary rocks and from 10^{-15} to 10^{-17} m^2 in the crystalline basement. As expected, the permeability distribution governs the dominant mode of heat transport, heat conduction or advection. It was not possible to calibrate this model, due to a lack of suitable data of adequate quality. However, a permeability value that is compatible with the temperature data could be determined within an order of magnitude for the sedimentary units and the crystalline basement using the temperature logs from the Bayreuth and Weiden boreholes and the KTB pilot borehole (Jobmann and Clauser 1994): $0.5 - 1.0 \times 10^{-14}$ m^2 for the sedimentary units and $0.25 - 5.0 \times 10^{-16}$ m^2 for the crystalline basement. This corresponds to predominant advective heat transport in the sediments and a minor advective contribution to the heat transport in the crystalline basement. Further fitting of the model to the specific conditions at the KTB site is not appropriate owing to the lack of detailed data at this scale. The temperature, heat flow density, and flow fields for this model are shown in Plate 13. The results of the Pe analyses (Table 3.2) are compared in Plate 14 with the Péclet numbers calculated with the model for various permeability distributions. It can be seen that simple approximation of the topography by two plateaus with different elevations is not sufficient for an interpretation of the empirical Péclet numbers determined from borehole temperatures. Only in the immediate vicinity of the Frankonian Line are the Péclet numbers calculated with the model in agreement with those empirical values. About 10 km from the Frankonian Line, however, the effect in the model caused by the sudden increase in elevation disappears, whereas the borehole data remains constant (Pe = -0.33 ± 0.2) within a 90 % confidence interval. This is considered by Jobmann and Clauser (1994) to be due to the influence of either the three-dimensional topography or of paleo-climatic signals in the analyzed T logs, causing an incorrect interpretation of the Pe analysis.

Kohl and Rybach (1995) studied the influence of the 3-D topography on the fluid flow field at both the regional and local scales in the area around the KTB borehole. The computations were made using the FRACTURE finite-element program (Kohl et al. 1993), in which the steady-state equations are solved for mutually coupled groundwater flow and heat transport in both porous and fractured media (Appendix 1). The area covered by the model for the regional simulation corresponds to that shown in Figure 3.1. In the 3-D model there is no fluid flow across the bottom and lateral boundaries. The hydraulic potential at the top of the model, which represents ground level, is held constant (Fig. 3.9). Permeability varies in this model from 10^{-18} m^2 for the crystalline basement to 10^{-16} m^2 for the sedimentary units (cf. Fig. 3.2). Figure 3.10 shows the resulting distribution of hydraulic potential, which reflects the 3-D topography of the area. In accordance with the hydraulic potentials, the flow fields are oriented predominantly N–S with a weaker E–W component.

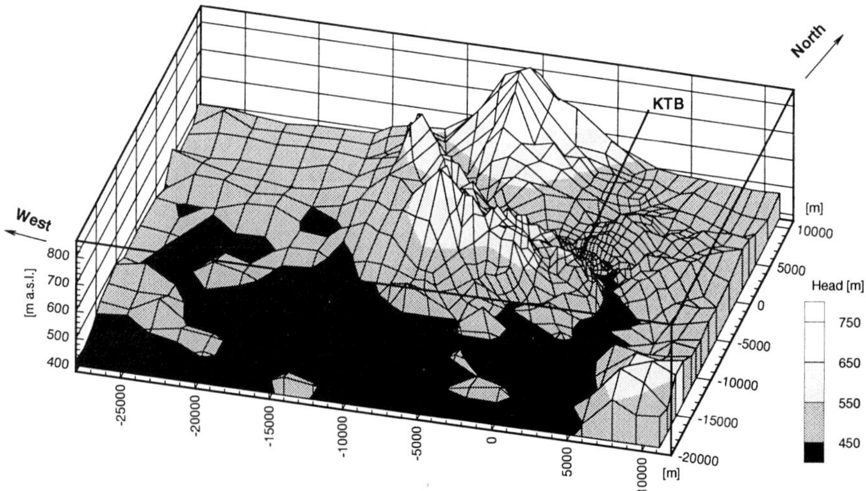

Fig. 3.9. Finite-element model of the topography in the KTB area (in m above m.s.l.), together with hydraulic potentials (in m) at the upper boundary of the regional hydraulic model of Kohl and Rybach (1995). The coordinates of the KTB site in the finite-element grid are (0, 0, 505).

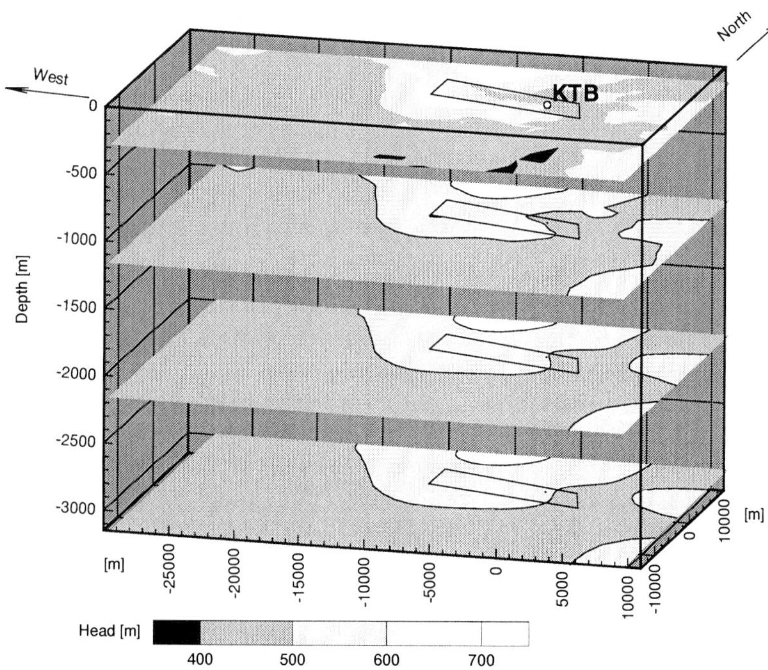

Fig. 3.10. Hydraulic potentials (in m) in the regional hydraulic model of Kohl and Rybach (1995). The outline indicates the location of the local model.

Coupled hydraulic and thermal computations were carried out for the area immediately surrounding the KTB site (the area of approximately 3 km × 12 km outlined in Fig. 3.10) using a local 3500 m thick (Fig. 3.11). The computations were made using a homogeneous permeability for the entire area, as well as only for the top 500 m of the model (Fig. 3.12). A fault zone was included in a variation of this model. The models are compared in Figure 3.13 with the data measured in the pilot borehole. It can be seen that the model for purely conductive heat transport (a) reproduces the empirical data rather well, in contrast to model (d), for which a homogeneous permeability of 10^{-15} m^2 was used. A further improvement of the advective models can be attained only by introducing a permeability structure below the top 500 m or by taking a fault zone into consideration (Kohl and Rybach 1995). Thus, the results of the 2-D simulation are in good agreement with the results of the 3-D models.

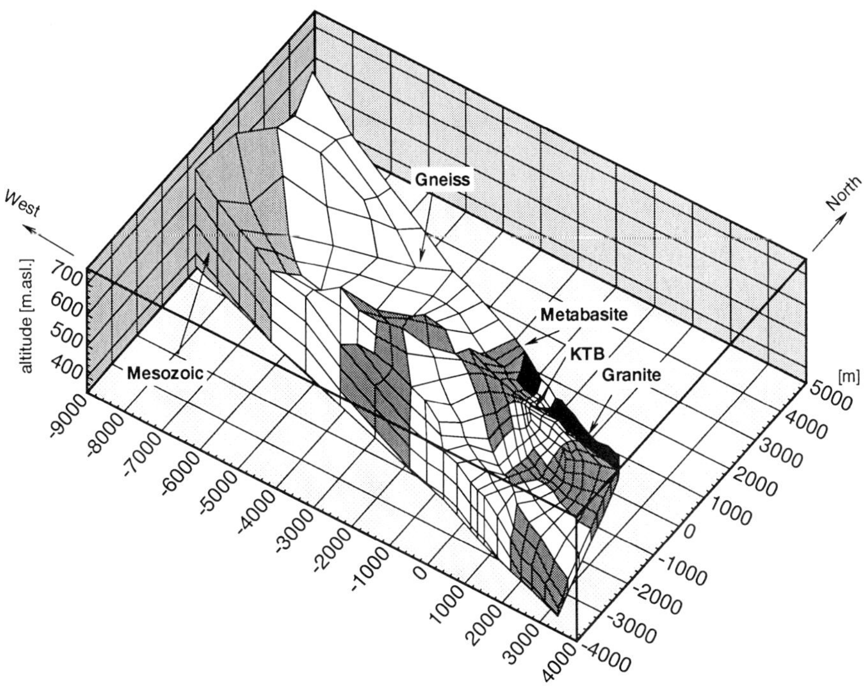

Fig. 3.11. Finite-element grid, lithological units from the surface to a depth of 350 m, and ground level (in m above m.s.l.) in the local model of Kohl and Rybach (1995).

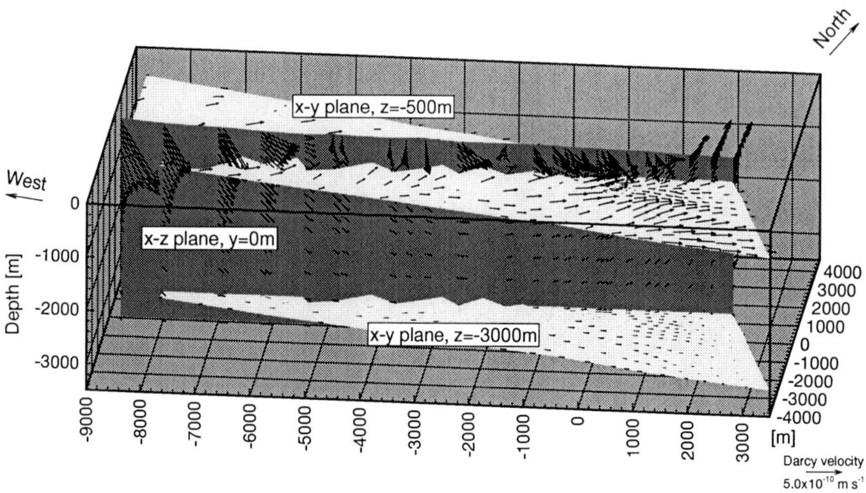

Fig. 3.12. Vertical and horizontal fluid flow fields for a homogeneous permeability of 10^{-15} m^2 in the local model of Kohl and Rybach (1995).

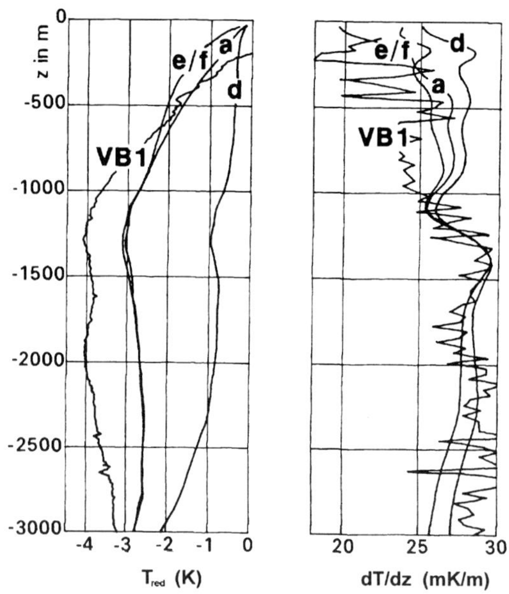

Fig. 3.13. Comparison of reduced temperature T_{red} and the vertical component dT/dz of the temperature gradient in the KTB pilot borehole with results of the local model of Kohl and Rybach (1995): a: pure conduction, d: homogeneous permeability, e: homogeneous permeability in the top 500 m, f: homogeneous permeability with an additional fault zone.

3.1.2
Is Heat Transport in the KTB Area Dominated by Advection Due to Free Convection in the Upper Crust or by Refraction of Conductive Heat Flow?

In contrast to expectations, the area of the KTB borehole comprises a considerable variety of steeply dipping geological units (Plate 12). Owing to differences in thermal conductivity of these units, especially between gneiss and metabasite, lateral conductive heat transport occurs in the area. To estimate the magnitude of this effect, both the local and regional situations were simulated. In the local 2-D model of Kohl and Rybach (1995) the geological units (Hirschmann 1993) (Plate 12) were simulated in considerable detail (Fig. 3.14). They include the effect of the anisotropy of the gneiss by distinguishing between steeply dipping, higher conductivity gneiss units and sub-horizontal, lower conductivity ones (Fig. 3.14, Table 3.4).

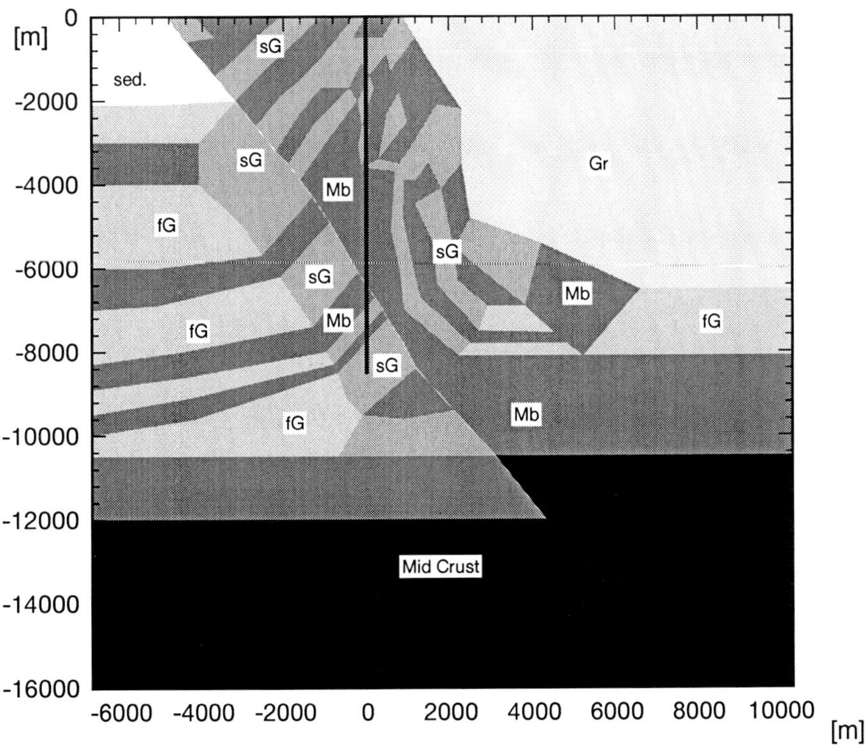

Fig. 3.14. Local 2-D model for simulating lateral conductive heat transport (Kohl and Rybach 1995) based on the geological model of the KTB area (Hirschmann 1993; Plate 12); see Table 3.4 for symbols.

Table 3.4. Thermal conductivity λ and heat production rate H in the local model (Fig. 3.14).

domain	λ [W m^{-1} K^{-1}]	H [μW m^{-1}]
Sediments (sed.)	2.0	0.4
Gneiss (sub-horizontally banded: fG)	3.0	1.5
(steeply banded: sG)	3.3	1.5
Metabasite (Mb)	2.5	0.8
Granite (Gr)	3.7	6.0
Mid-Crust	3.4	0.6 – 1.0

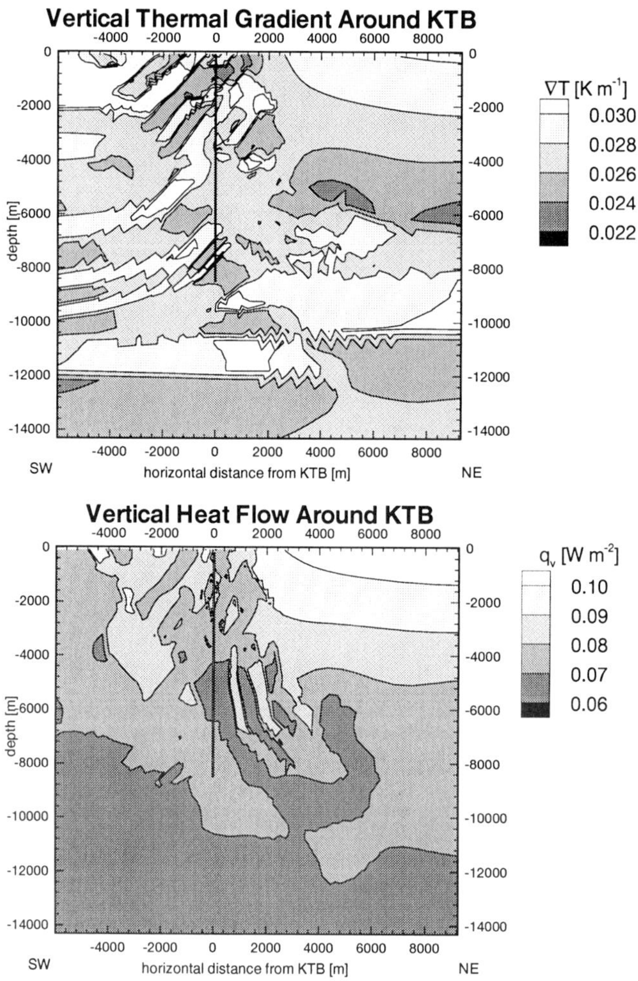

Fig. 3.15. Vertical components of the temperature gradient and the heat flow density in the local 2-D model (Kohl and Rybach 1995; see Fig. 3.14 and Table 3.4).

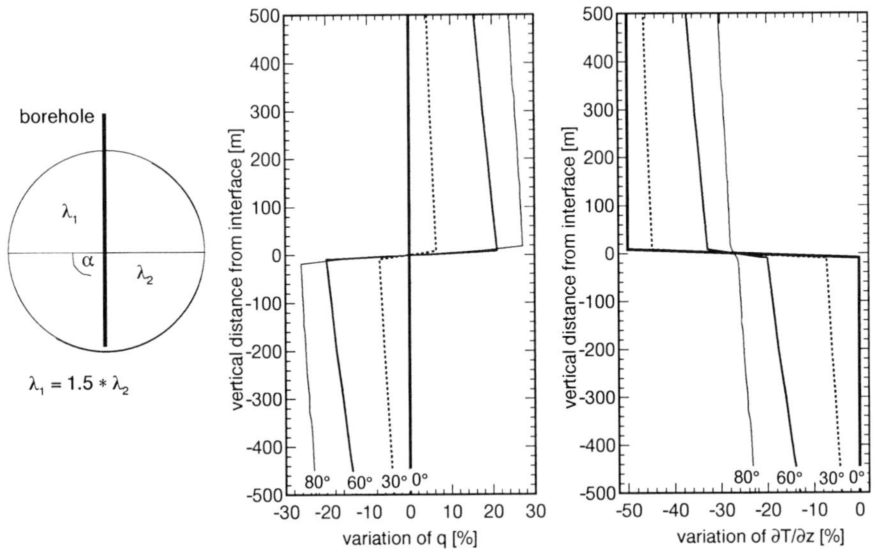

Fig. 3.16. Plot of the vertical components of heat flow density (q) and the temperature gradient ($\partial T/\partial z$) as a function of the angle α between the horizontal and the boundary between two media of different thermal conductivities, λ_1 and λ_2 (Kohl and Rybach 1995).

In these simulations the lateral boundaries were impermeable to heat and fluid flow, and the temperature at the surface was fixed at a constant 8 °C. Heat flow density was varied at the base of the model to optimize the results. A value of 63 mW m^{-2} (corresponding to 437 °C at a depth of 16 km) yielded the best agreement with the temperature data from the pilot and main KTB boreholes. The maximum deviations were less than ± 2.5 K. The vertical temperature gradient and vertical heat flow density field obtained with this model are shown in Figure 3.15. It can be seen that the vertical component of the temperature gradient is constant over a depth interval of about 2-6 km in the area of the steeply dipping layers immediately around the KTB site. In contrast, the vertical component of the heat flow density in this area varies considerably. This is in good agreement with the temperature gradients in the KTB boreholes (Plates 2 and 3), where it can be seen also that the heat flow density correlates with the thermal conductivity. These three observations are a clear indication of lateral conductive heat transport resulting from the lateral differences in thermal conductivity. This is clearly confirmed by the model calculations of Kohl and Rybach (1995).

The reason for the difference in the signals in the temperature gradient and in the heat flow density is the difference in the continuity of their vertical components: The vertical component of the heat flow density is continuous at horizontal thermal conductivity boundaries; that of the temperature gradient is discontinuous. In the vicinity of vertical thermal conductivity boundaries, the situation is just the opposite: The vertical component of the temperature gradient is continuous and that of the heat flow density is discontinuous. This relationship, which is important for understanding heat transport at the KTB site, is illustrated in Figure 3.16: When the dip is between 60° – 80°, the variation in the vertical temperature gradient is distinctly less than that of the vertical heat flow density. This behavior is seen, for example, in the main KTB borehole between depths of 2 and 8 km.

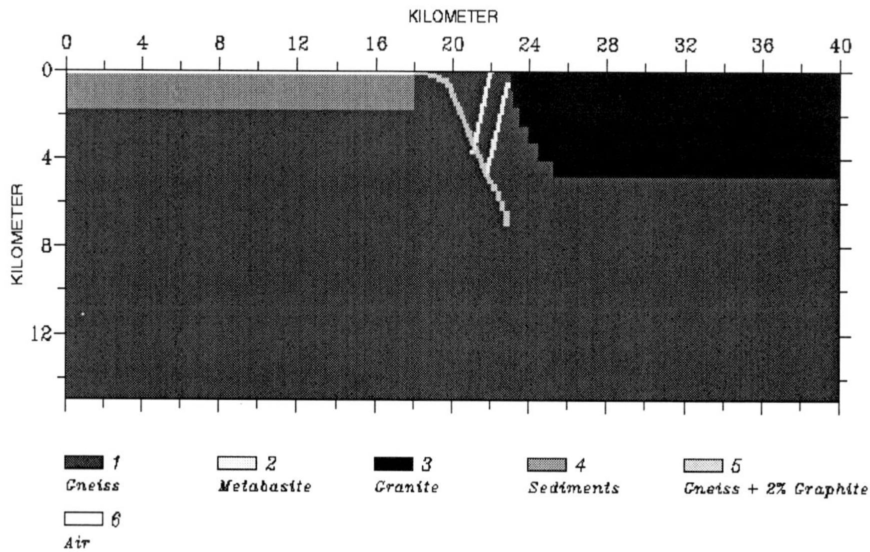

Fig. 3.17. Vertical, 2-D, conceptual model for simulating regional heat transport in the upper crust (Jobmann and Clauser 1994) along profile A—B (Fig. 3.2) perpendicular to the Frankonian Line (KTB: km 22): The geological units are highly simplified after Hirschmann (1993).

Table 3.5. Parameter distribution in the 2-D regional model down to 15 km (Fig. 3.17) and in the version of this model extended to a depth of 30 km (Plate 17): thermal conductivity λ, heat production rate H, porosity Φ and permeability k.

	domain	λ [W m^{-1} K^{-1}]	H [μW m^{-3}]	Φ [%]	k [10^{-15} m^2]
1:	gneiss	3.3	<15 km: 1.5 >15 km: 0.8	1	< 4.75 km: 0.25, 0.025 > 4.75 km: 0.001
2:	metabasite	2.6	0.6	5	0.5, 0.05
3:	granite	3.0	6.0	3	0.25, 0.025
4:	sediments	2.2	0.4	15	10
5.:	gneiss containing 2% graphite	4.5	1.5	5	0.1
6:	air	0.0	0.0	—	0

Jobmann and Clauser (1994) and Lehmann et al. (1998) investigated the influence of both conductive and advective lateral heat transport at a regional scale using a vertical 2-D model of the upper crust (Table 3.5, Fig. 3.17) along profile A–B perpendicular to the Frankonian Line (Fig. 3.2). The lateral differences in thermal conductivity in this model are due mainly to: (a) The graphitic rocks observed near the Franconian Line, which are modeled as permeable, graphitic gneiss owing to its history of high tectonic stress; (b) The two metabasite bodies that dip in the opposite direction of the contact between the Mesozoic sediments and the gneiss; (c) The granite pluton. The considerable influence of even low graphite concentrations on thermal conductivity is illustrated in Figure 3.18 on the basis of measurements on samples from the Kropfmühl graphite occurrences near Passau, southeast Germany: As little as 5 % graphite causes a doubling of the thermal conductivity.

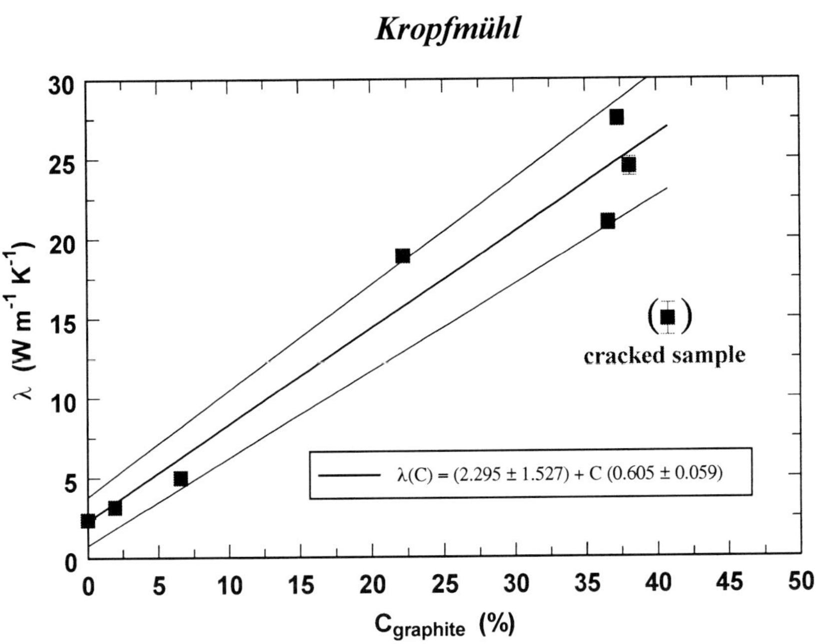

Fig. 3.18. Thermal conductivity λ of gneiss as a function of graphite content C.

The boundary conditions are the same as in the model for shallow depths illustrated in Figure 3.8 except for the heat flow density at the base of the model at a depth of 15 km, whose magnitude was varied to optimize the results. A value of 45 mW m^{-2} yielded the best agreement with the temperature data from the pilot and main KTB boreholes. Thermally relevant free convection can develop only if the permeability in the mid-crust is sufficiently large: Free convection occurs in the model throughout the upper crust at permeabilities above about 10^{-17} m^2. This is in addition to the forced convection system resulting from the topography (Plate 15). If the permeability of the crust below a depth of 4.75 km is reduced from 2.5×10^{-17} m^2 to 10^{-18} m^2, free convection no longer occurs to a significant extent and no longer significantly affects heat transport (Plate 16). Therefore, a threshold permeability value of 10^{-17} m^2 can be defined for this system, above which free convection can be expected throughout the upper crust. Interestingly, analysis of pumping tests at the final depth of the KTB borehole yielded permeabilities on the order of magnitude of this threshold value of 10^{-17} m^2 (Huenges et al. 1997). The fit of this model to the KTB data is illustrated in Plate 18. It cannot, of course, be as good in this regional conceptual model as in the detailed, local model presented above. It yields temperatures that are too high for great depths. In contrast, down to about 6 km the values from the advective model are too low, whereas below that they are closer to the measured values. This is a clear indication of the significance of the distribution of heat production rate in the mid-crust for the model results.

For this reason, the model described above was extended to greater depth. This model represents an extension of the regional model in Figure 3.17 to a depth of 30 km and contains a variation in heat production rate with depth between about 3-1 μW m^{-3} in the mid to lower crust (Clauser et al. 1997). Optimization of the heat flow density at the base of the model with respect to the measured temperatures yielded a value of 28 mW m^{-2}. Like for the distribution of the heat production rate, the model results are very sensitive to the value of the heat flow density at the base: a change from 30 to 33 mW m^{-2} yields a temperature increase of 45 K at the base of the model. The computations were carried out with a simulation program (Wang and Beck 1989) adapted to the conditions of the KTB area by Lehmann et al. (1998). Temperature, heat flow density and Darcy velocity obtained in this simulation are shown in Plate 17. It can be seen that by varying heat production rate with depth a distinct improvement in the agreement of the model with the measured data is obtained that is comparable to that obtained with the local, detailed model (Plate 18).

In summary, the following statements can be made about heat transport at the KTB site: (1) The near-surface temperature field is affected by local fluctuations in topography and the associated groundwater flow; (2) Temperature, temperature gradient and heat flow density in the top 10 km are characterized by lateral conductive heat transport; (3) The distribution of radiogenic heat sources has a decisive influence on the temperature regime and heat transport in the mid- to lower crust; (4) Hydraulic tests indicate permeabilities in the crystalline basement close to 10^{-17} m^2, a threshold value for the onset of thermally effective, free convection throughout the crust. (5) Processes on a regional scale are simulated well by 2-D models. Analysis of the local situation, however, requires a 3-D model for this complicated geology. This normally presumes, however, the availability of three-dimensional data about the distribution of the parameter values. If such information is not available, then two-dimensional models may also provide insight if the simulated structures approach two-dimensionality.

3.2 Kola Peninsula: Steady-State and Transient Heat Transport Processes

The SG-3 borehole on the northern rim of the Kola Peninsula (69°24'N, 30°35'E), which reached a depth of 12,261 km in 1991, is still the deepest borehole in the world (Fig. 3.19). The cores and borehole logs (Kozlovsky 1987; Kremenetsky and Krivtsov 1991) yielded considerable new information, but at the same time also raised a number of new questions. With respect to heat transport in basement rocks, the main results were (a) a heat flow density that varied with depth (Kremenetsky and Ovchinnikov 1986a,b) and (b) porosity and permeability values that had not been expected at such depths (Bayuk et al. 1987). In the following section, therefore, a simplified model will be used to determine the influence of steady-state heat advection and lateral heat conduction on heat transport in the area of the Kola borehole. The downward diffusion of paleo-climatic signals will also be modeled. The results of the two models will then be compared to determine how well they explain the observed data.

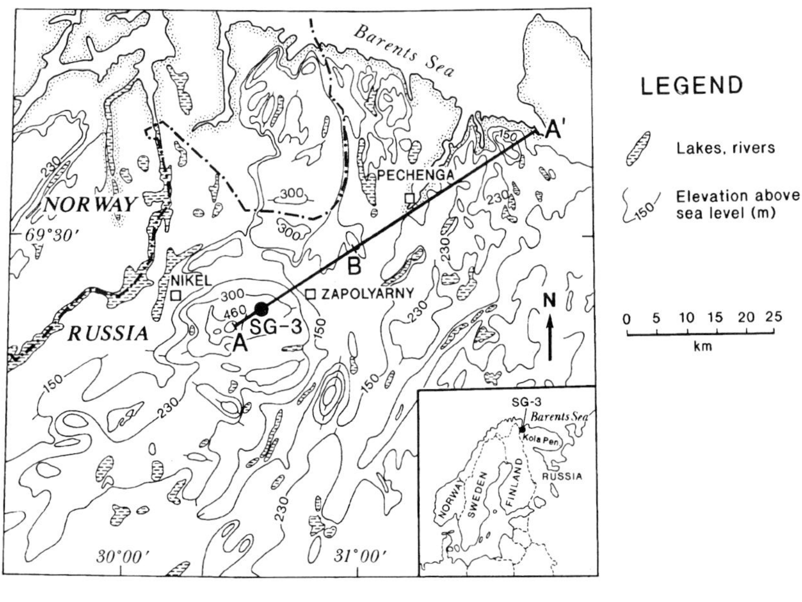

Fig. 3.19. Topographic map of the area around the Kola SG-3 borehole (simplified after US Air Force, 1971); elevations in m.

3.2.1 Does Heat Transport Occur Under Steady-State Conditions by Advection Resulting from Forced Convection or by Refraction of Conductive Heat Flow?

The Kola borehole was drilled in a hilly, glacial landscape with a maximum elevation of 700 m just south of the SG-3 borehole, falling off to the NE to the Barents Sea. The elevation varies between about 150 and 500 m (Fig. 3.19).

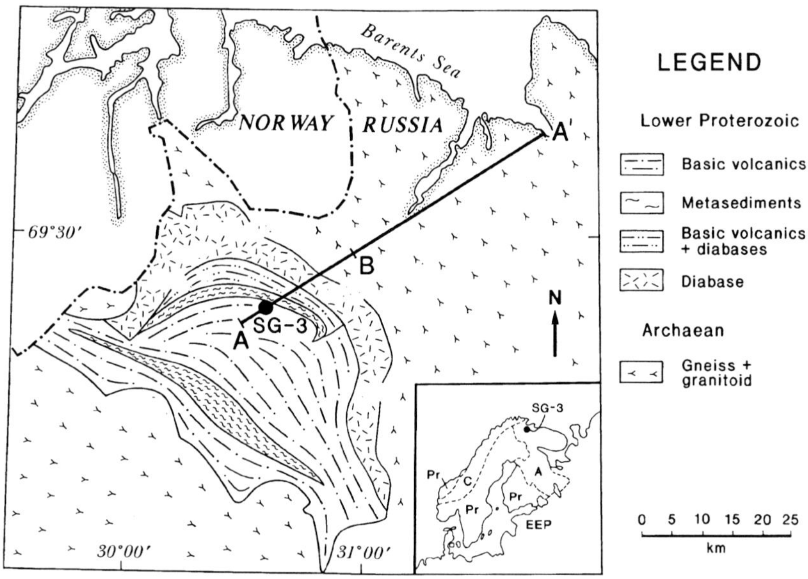

Fig. 3.20. Geological map of the area around the Kola SG-3 borehole (simplified after Nalivkina et al. 1987); A: Archean, Pr: Proterozoic, C: Caledonian, EEP: East European Platform.

Geologically, the borehole is in the center of the ore-rich, Pechenga structure in an Archean shield, a syncline whose limbs both dip SW (Fig. 3.20). For this specific location, neither the topography nor the geology can, strictly speaking, be simplified to a 2-D model. For a 3-D model, however, it is necessary for the parameter distributions of the model to be known in three dimensions. Since this is not the case for the Kola region, 2-D conceptual models must be used to better understand the relative influences of the different heat transport mechanisms. Similar to the situation at the KTB site, the main heat transport mechanisms (besides vertical heat conduction) are lateral heat conduction, caused by differences in thermal conductivity and dip of the layers, and advection driven by a difference in elevation. The influence of these two processes is first simulated in a 2-D conceptual model. A vertical, regional, 2-D model along profile A—A' (shown in Figs. 3.19 and 3.20) was set up for this purpose. This simplified model was prepared on the basis of the available NE–SW geological sections (Nalivkina et al. 1987; Lanev et al. 1987; Kazansky et al. 1987), including the most important units, but leaving out details (Fig. 3.21).

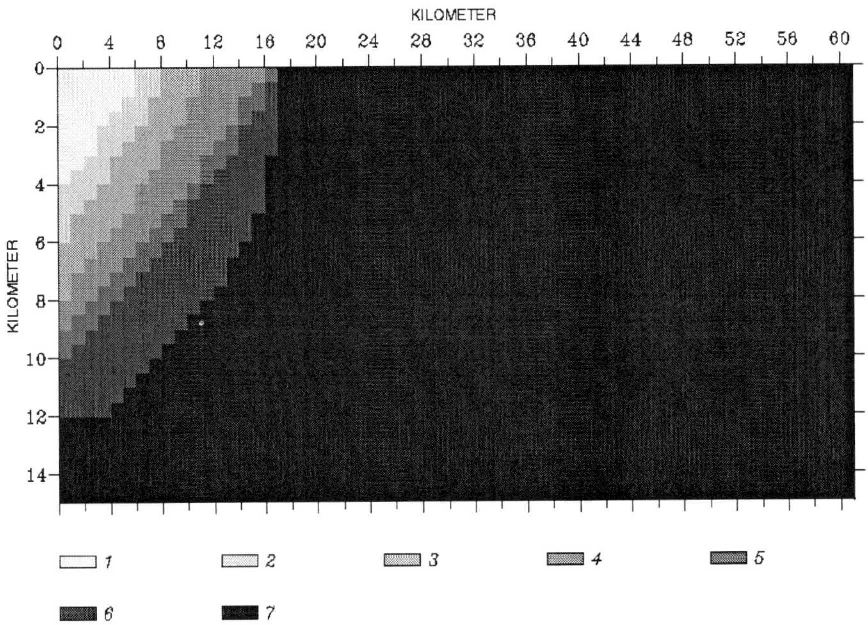

Fig. 3.21. Vertical, 2-D, conceptual model for simulating heat transport in the upper crust (Kukkonen and Clauser 1994) along profile A—A' (Fig. 3.19) on the Kola Peninsula (SG-3: km 5); see Tables 3.6 and 3.7 for the parameter values.

Table 3.6. Parameter distribution of the rocks of the Pechenga structure modeled in Fig. 3.21 (λ: thermal conductivity, H: heat production rate).

domain	age	λ [W m^{-1} K^{-1}]	H [μW m^{-3}]
1: basic volcanics 1	Proterozoic	2.5	0.3
2: metasediments	Proterozoic	3.3	1.4
3: basic volcanics 2	Proterozoic	3.5	0.3
4: diabase	Proterozoic	2.4	0.7
5: Archean gneiss 1	Archean	2.4	1.2
6: Archean gneiss 2	Archean	2.2	1.0
7: Archean gneiss 3	Archean	3.0	1.0

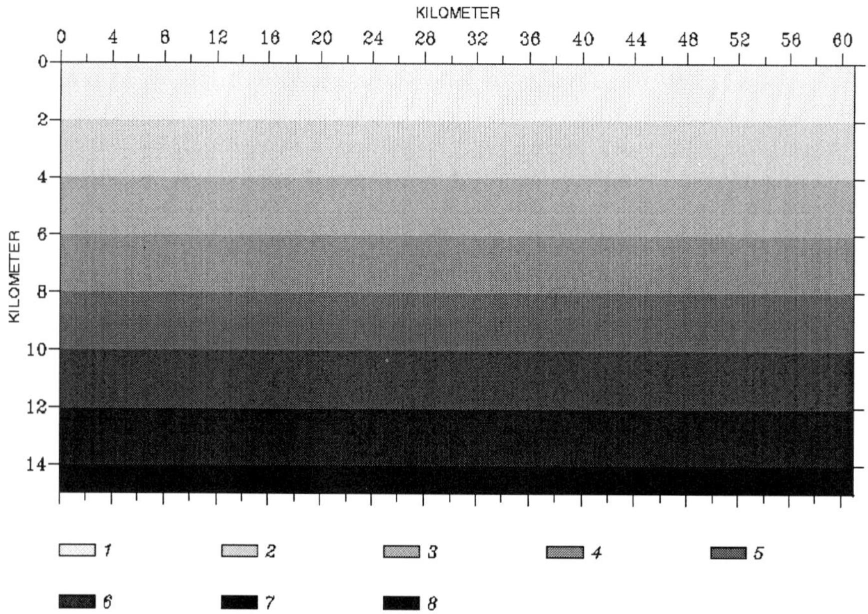

Fig. 3.22. Permeability distribution in the 2-D conceptual model (Fig. 3.21; SG-3: km 5). For values of permeability see Table 3.7.

Table 3.7. Permeability k in the 2-D conceptual model (Fig. 3.22; SG-3: km 5)

domain	k $[m^2]$
1	8×10^{-16}
2	2×10^{-16}
3	1×10^{-16}
4	1×10^{-17}
5	1×10^{-18}
6	1×10^{-19}
7	1×10^{-20}
8	1×10^{-21}

Thermal conductivities (Kremenetsky and Ovchinnikov 1986a,b; Galdin et al. 1986, 1987) and heat production rates (Arhavskaya et al. 1987; Kremenetsky et al. 1989) were selected on the basis of the lithology (Table 3.6). A decrease of one order of magnitude per 2 km depth was assumed for the permeability owing to the decrease in fracture frequency with depth (Borevsky et al. 1987) (Fig. 3.22). A transmissivity of 1 – 2 m^2 d^{-1} was reported by Borevsky et al. (1987) for the top 2 km, which corresponds to a permeability of about 10^{-15} m^2. Bayuk et al. (1987) measured permeabilities of samples several cm across. They are not sufficiently representative, however, to correlate them with specific lithologies or to extrapolate them from the centimeter scale of the samples to the kilometer scale of the model (cf. Clauser 1992). The porosity values determined on the samples were mostly smaller than 1 %, with maximum values of 5 % (Bayuk et al. 1987). Therefore, a constant porosity of 1 % was assumed in the model.

The grid spacing of the model was 500 m × 1000 m. A constant temperature of 1 °C was used as boundary condition at the top of the model, and the hydraulic potentials appropriate for the elevations were used along profile A—A' (Fig. 3.19). The topography can be simulated more precisely using the model with a finer grid described in section 3.2.2. No-flow boundary conditions apply at the sides of the model for fluid and heat flow. The boundary conditions at the lower edge of the model are a constant heat flow density of 40 – 50 mW m^{-2} and no fluid flow across the boundary (Table 3.8). Plate 19 shows a typical result of a simulation. The relatively large variability of the vertical component of heat flow density in the top 5 km of the model correlates strongly with the local, topography-driven flow systems.

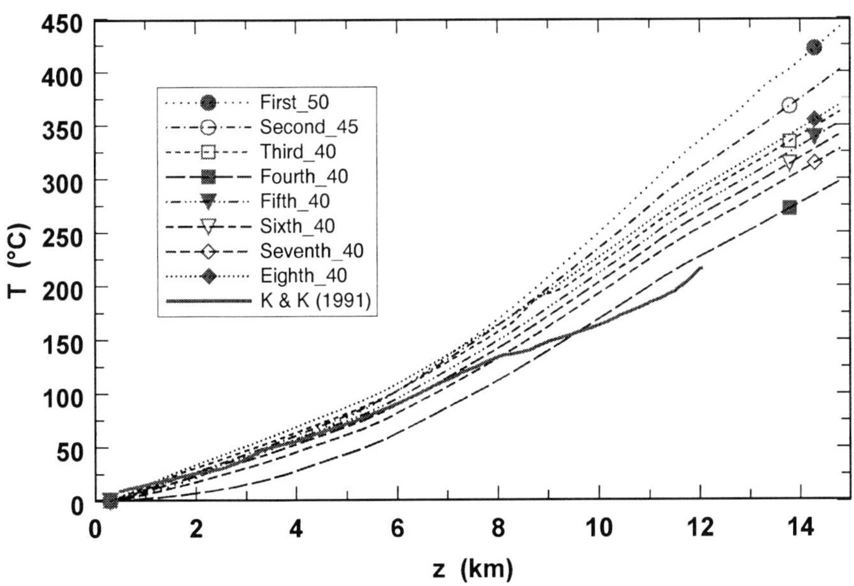

Fig. 3.23. Comparison of the steady-state simulation results with the temperature profile published for the SG-3 borehole (Kremenetsky and Krivtsov 1991). For the parameter values of the model, see Table 3.8.

The available data is insufficient for a model calibration, but for the location of the super-deep SG-3 borehole the simulation results can be compared with temperature and heat flow density data published by Kremenetsky and Krivtsov (1991) and Kremenetsky and Ovchinnikov (1986b), respectively. Because drilling operations have been going on at the SG-3 borehole, with intermissions, for more than 20 years, the temperature profile below 3 km depth shows clear indications of thermal transients which were most probably caused by technical operations. Because it had been recorded after a shut-in of the borehole that had lasted for 1 ½ years, the T log of Kremenetsky and Krivtsov (1991) above a depth of 7.2 km probably reflects the undisturbed formation temperatures within a range of ± 5 K. No estimate of accuracy can be given for the part of this T log below this depth. However, the temperatures recorded in that part of the borehole are, without a doubt, lower than the true undisturbed formation temperature (Kukkonen and Clauser 1994).

The simulation results of the different models are compared for this temperature profile in Figure 3.23. The models differ with respect to the permeability and heat flow density values selected at the base of the models (Table 3.8). Above a depth of 7.2 km, the results of a number of the models are in agreement with the empirical temperature profile of the SG-3 borehole within ±10 K. In view of the precision of this profile discussed above, no attempt was made further improve the model fit by more parameter variations.

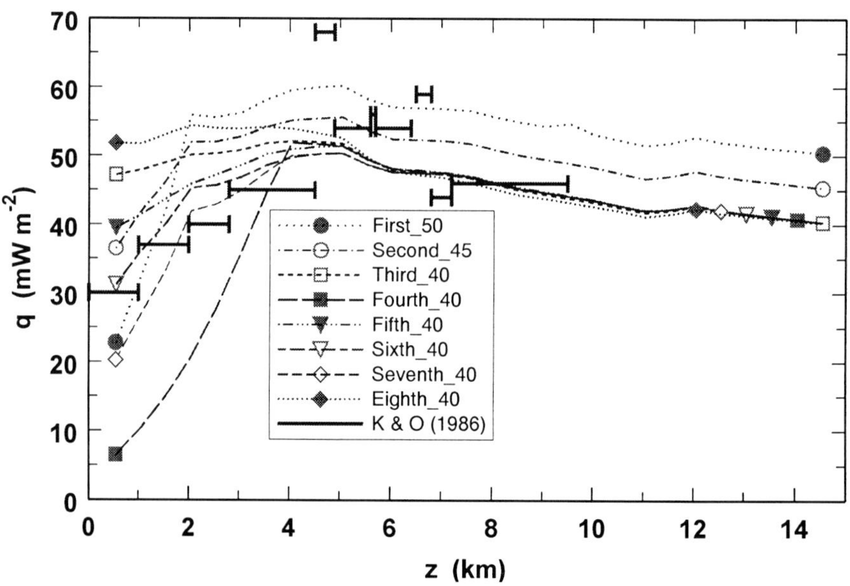

Fig. 3.24. Comparison of the steady-state simulation results with the heat flow density data published for the SG-3 borehole (Kremenetsky and Ovchinnikov 1986b). For the parameter values of the model, see Table 3.8.

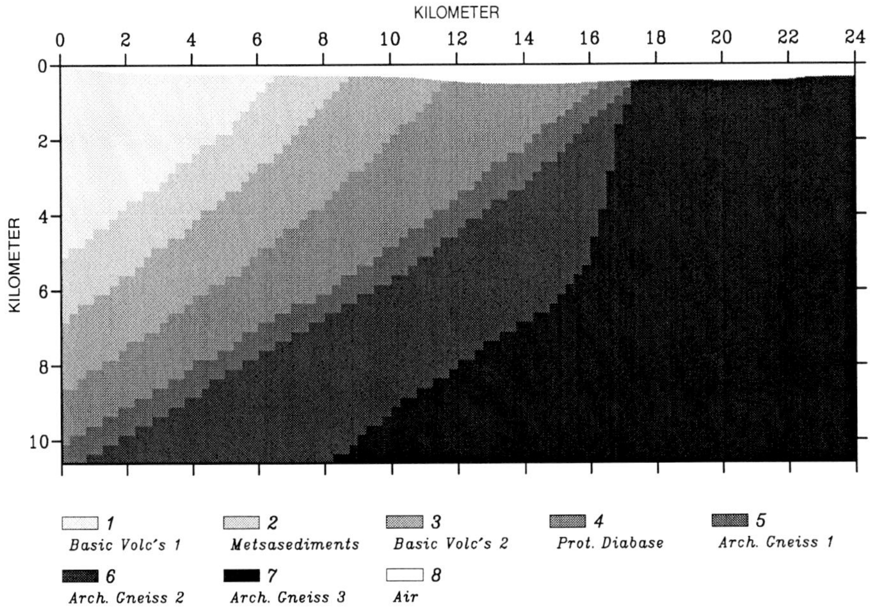

Fig. 3.25. A further refinement of the conceptual model shown in Fig. 3.21 for simulating heat transport in the upper crust (Kukkonen and Clauser 1994) along profile A—B (Fig. 3.19) on the Kola Peninsula (SG-3: km 5). For the parameter values of the model, see Tables 3.6 and 3.7.

Table 3.8. Heat flow density q_B in mW m^{-2} at the base and permeability k in m^2 between 0 km and 2 km and between 2 km and 4 km in the conceptual models (Fig. 3.21 and Fig. 3.25). Below 4 km the permeability decreases in all models: from 10^{-16} m^2 at a depth of 4 – 6 km to 10^{-21} m^2 at a depth of 14 – 15 km (with the exception of the homogeneous model 'EIGHTH_40').

model name	q_B [mW m^{-2}]	$k_{0\text{-}2\ km}$ [m^2]	$k_{2\text{-}4\ km}$ [m^2]
FIRST_50	50	1×10^{-14}	1×10^{-15}
SECOND_45	45	5×10^{-15}	5×10^{-16}
THIRD_40	40	1×10^{-15}	1×10^{-16}
FOURTH_40	40	1×10^{-14}	1×10^{-14}
FIFTH_40	40	1×10^{-15}	1×10^{-15}
SIXTH_40	40	5×10^{-15}	1×10^{-15}
SEVENTH_40	40	8×10^{-15}	1×10^{-15}
EIGHTH_40	40	1×10^{-21}	1×10^{-21}

The corresponding comparison for heat flow density is shown in Figure 3.24. This parameter is more sensitive to differences between the models and provides an important constraint on the heat flow density at the base of the model and the permeability values. For example, values in the range of 29 – 33 mW m^{-2} for heat flow density at the surface, measured in the SG-3 borehole and some of the neighboring boreholes, are obtained only with some of the models. The results thus indicate permeabilities of up to 10^{-15} m^2 for the top 4 km and a heat flow density of 40 mW m^{-2} at the base of the model, such as, for instance, in model 'SEVENTH_40' (Plate 19, Table 3.8).

3.2.2 How Much Do Paleo-Temperatures at the Earth's Surface Influence the Thermal Regime at Depth?

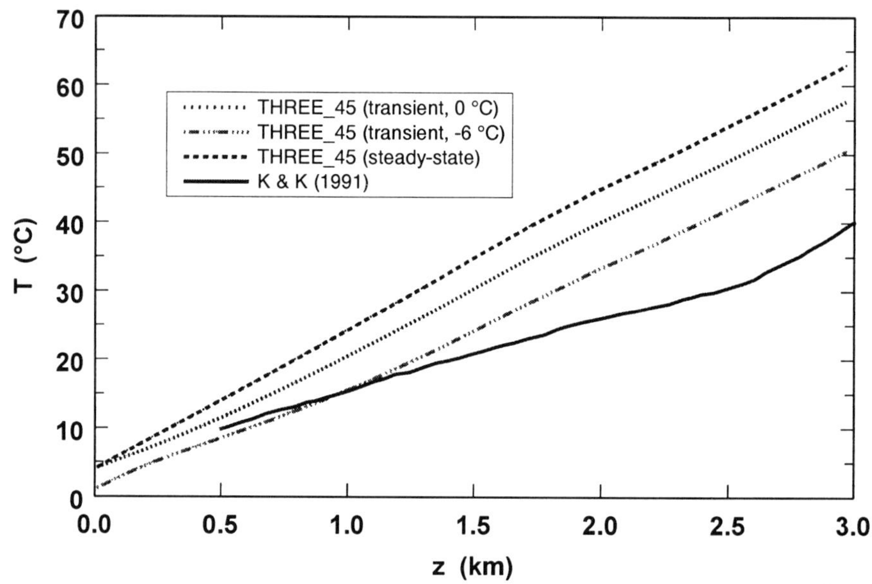

Fig. 3.26. Comparison of the transient simulation results with the temperature profile published for the SG-3 borehole (Kremenetsky and Krivtsov 1991). Assumed ground temperatures during the Ice Ages are given in parentheses.

Owing to the rapid fading of paleo-climatic signals with depth (Plate 1), the simulation of these effects required a more refined grid, especially the upper part of the model. For the transient simulations, a more detailed model of profile A–B (Figs. 3.19 and 3.20) was therefore prepared with a grid spacing of 20 m × 250 m for the top 160 m and 250 m × 250 m below that depth (Fig. 3.25). Constant temperatures and hydraulic potentials were fixed at the upper boundary condition. The closely spaced grid of this model makes it possible to better simulate the topography, so that the constant hydraulic potential in the model matches the actual ground surface. No-flow boundary conditions apply again at the lateral boundaries. The lower boundary of the model has a heat flow density fixed at 45 mW m^{-2} and is impermeable to fluid flow.

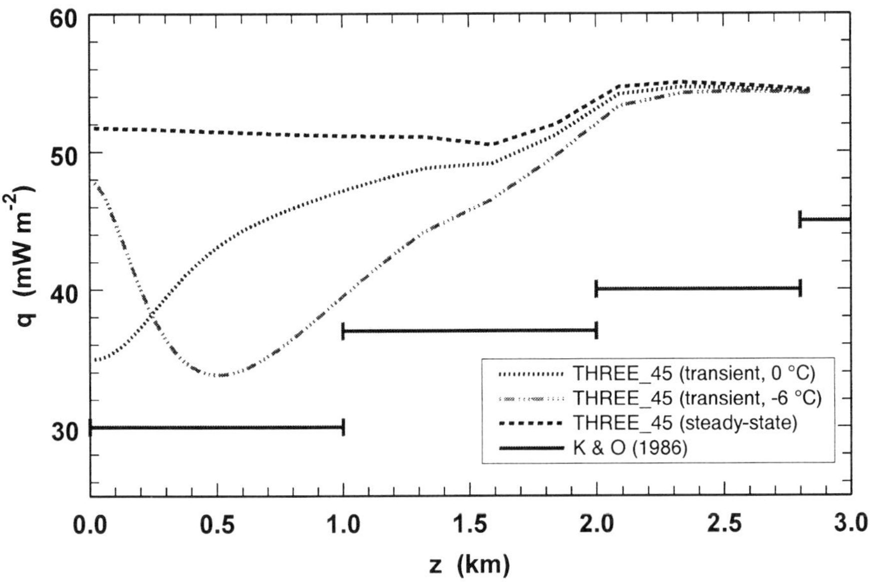

Fig. 3.27. Comparison of the transient simulation results with the heat flow density published for the SG-3 borehole (Kremenetsky and Ovchinnikov 1986b). Assumed ground temperatures during the Ice Ages are given in parentheses.

This more detailed model was used to simulate the downwards diffusion of transient temperatures from the Earth's surface. The thermal diffusivity κ was calculated from the thermal conductivity λ as $\kappa = \lambda / \rho c$, using an assumed constant thermal capacity of ρc = 2.3 MJ m^{-3}. The time-dependence of the temperatures at the upper boundary of the model was taken into consideration by assuming three different values for temperature at the top boundary during three different periods: (1) Either 0 °C or -6 °C during the glaciation up to 10,000 years BP; (2) A temperature 2 K higher than at present during the post-glacial period from 10,000 to 1,000 years BP; (3) The present temperatures for the last 1000 years. For the Ice Age either a warm temperature (0 °C) or a cold temperature (-6 °C) was assumed for the bottom of the ice sheet. Temperatures of 0 °C at the bottom of the ice sheet are supported by model calculations of the growth and advance of the ice sheets from Scandinavia to northern Germany between 50,000 – 10,000 years BP (Delisle 1991). A considerably cooler value of -6 °C was also used for comparison. During the two post-glacial periods, the surface temperatures in the model follow an adiabatic temperature gradient in air of 6 mK m^{-1}, corresponding to the present elevation. For comparison, a steady-state computation with the present surface temperatures was also carried out.

The modeling results are compared in Figures 3.26 and 3.27 with the temperature and heat flow density data from the SG-3 borehole. The paleo-climatic signal has largely faded below a depth of 1.5 km. This is in good agreement with the discussion in chapter 2 (Plate 1 and Fig. 2.2). In the present case, the paleo-climatic signal is too small to explain the observed temperature and heat flow density signals, even assuming an extreme temperature of -6 °C at the base of the ice sheet. It is, therefore, concluded that the observed differences in heat flow density in the SG-3 borehole are caused mainly by advective heat transport and refraction of the conductive heat flow owing to contrasts in thermal conductivity and to the dip of the strata.

4 Recognizing Deep Underground Processes: Influence of an Intrusion in the Lower Crust on the Thermal Regime in the Upper Crust

The analysis of temperature and heat flow density fields is often used to identify processes deep underground. The applications range from geodynamic problems on a global scale (e.g., Pollack et al. 1993; Stein 1995) to the quantification of local geothermal resources (e.g., Hänel and Staroste 1988). All of these applications are based on the fact that, owing to the temperature difference between the interior of the Earth and its surface, thermal signals at depth eventually reach the Earth's surface with the terrestrial heat flow. As discussed in the previous chapters, these signals can be superimposed, however, by a number of other processes on their way to the Earth's surface. Therefore, their identification and interpretation are not trivial problems.

The temperature field and heat flow density at the Earth's surface, usually determined in boreholes up to several hundred meters deep (Chapman et al. 1984), are generally the starting point for such studies. The question, therefore, is how readily can thermal events at depth be recognized in such data. This will now be analyzed using a concrete case as an example: an intrusion in the lower crust.

The geological situation on which the model is based is shown in Figure 4.1. This situation is typical of conditions in continental graben systems: A granitic pluton intrudes into the lower crust, adding heat and water of crystallization to the crust. The 2-D conceptual model is shown in Figure 4.2. The simulation model comprises the top 15 km of the crust. At the base of this model the intrusion causes variations in space and time of the boundary conditions for heat and fluid flow. The heat input from the intrusion is taken from the results of thermo-mechanical modeling of a granitic intrusion in the lower crust (Kukowski and Neugebauer 1990). The thermal and hydraulic response in the upper crust to this time- and location-dependent lower boundary condition is then analyzed. The upper crust is modeled as a porous medium, using the SHEMAT program described in section 3.1.1.2. This program is used to solve the equations for density-dependent flow, as well as diffusive and advective heat transport (Appendix 1). The computations take into account both the temperature dependence of the thermal conductivity of the rock and the temperature and pressure dependence of density, viscosity, specific heat capacity and thermal conductivity of water.

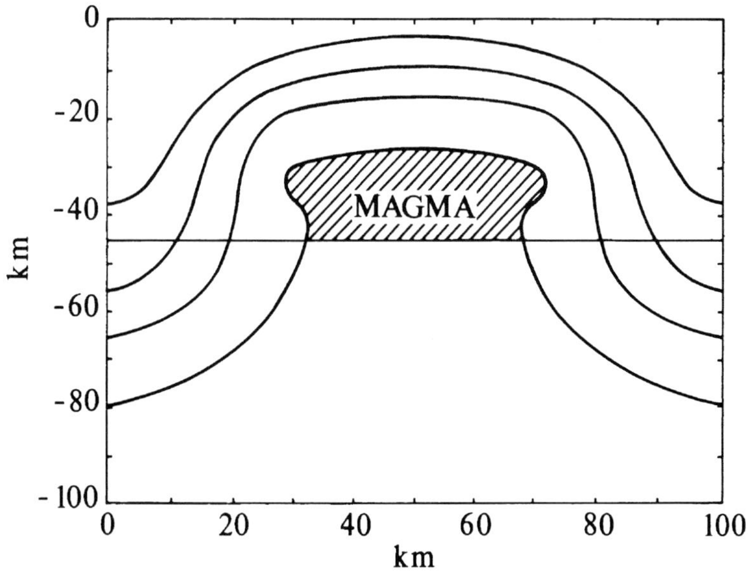

Fig. 4.1. Emplacement of a pluton in the lower crust leads to a transient updoming of the isotherms. The conceptual model in Fig. 4.2 represents the top 45 km of this figure.

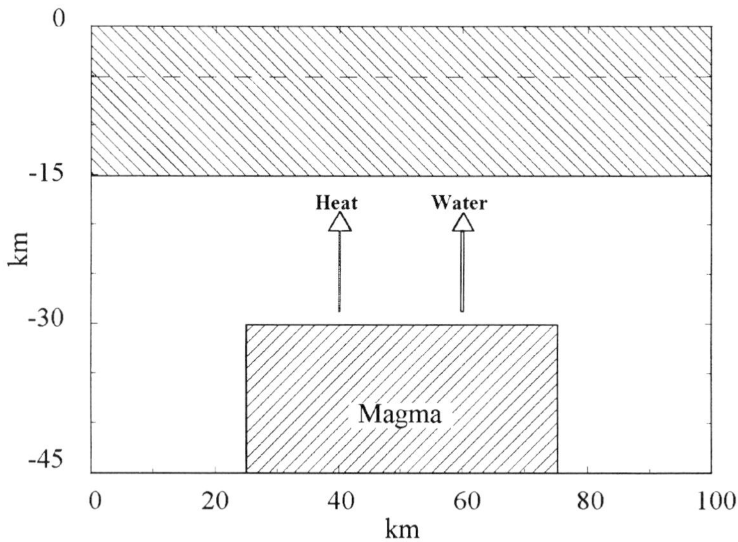

Fig. 4.2. Conceptual model of the upper half of Fig. 4.1. Fluid flow and heat transport are simulated for the top 15 km of the crust, which is modeled as a porous medium .

For numerical stability and to minimize numerical dispersion, a fully implicit time-discretization procedure was selected for solving the flow and transport equations (Appendix, equations 29 and 30). The advection term of the transport equation was weighted using Ilyin-type upstream weighting, which depends on the local Darcy velocity (Clauser and Kiesner 1987). Differences of 1 µm and 1 µK in hydraulic potential and temperature between two successive iterations were used as convergence criteria for the iterative solution of the flow and heat transport equations. The finite-difference computations used a block-centered 100×15 grid with a uniform horizontal and vertical grid spacing of 1 km. The computations were normally carried out for a time period of 30 Ma, after which the thermal anomaly under purely conductive conditions has faded. Considerably longer simulation periods of up to 1000 Ma were necessary for models with a large advection component.

Owing to the non-linearity caused by the mutual coupling of the flow and heat transport equations, the time intervals can no more be selected arbitrarily, in spite of the fully implicit formulation used. For this reason, the time intervals were kept constant or varied during the computations, depending on the strength of the convection system. Depending on the model, the constant time intervals were set to a value between 500 a and 10,000 a; the variable time intervals varied between 0.1 Ma and 2.5 Ma. The duration of the variable time steps were calculated automatically according to the maximum temperature difference in respect to the previous time step: If this difference was less than 10 mK, the length of the following time interval was doubled. If it was larger than 10 K, the computations were repeated with a time interval half as long.

The initial conditions for all computations were a hydrostatic pressure distribution and a temperature field corresponding in most of the model to a temperature gradient of 30 mK m^{-1}. At the top boundary of the model the initial temperature was 11 °C; at the bottom boundary it was set to 200 °C. The results of the computations changed little with changes in the initial conditions, as temperature and pressure adapt very quickly to the imposed boundary conditions. Constant hydraulic potential and temperature is specified for the top of the model, no fluid and heat flow across the lateral boundaries. In one model, a mass flow density m_B of 0 – 100 g a^{-1} m^{-2} is specified at the bottom of the model in order to simulate the dehydration of magma at depth as it cools. In all other models, the base of the model is impermeable in respect to flow, and the heat flow density q_B is varied between 20 and 40 mW m^{-2} as a function of location and time. The fading of the thermal anomaly associated with the intrusion at a depth of 15 km and its approximation by a boxcar and a staircase function are shown in Figure 4.3. The lateral variation of these functions at the base of the model is shown in Figure 4.4. Both are derived from the results of thermo-mechanical modeling of a mantle intrusion by Kukowski and Neugebauer (1990).

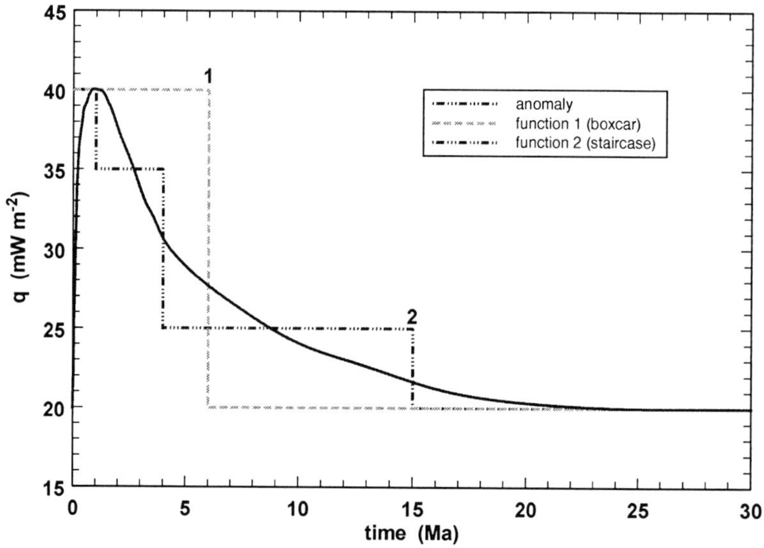

Fig. 4.3. Fading of a thermal anomaly at a depth of about 15 km, caused by emplacement of a pluton in the lower crust, and the approximation of this curve by two step functions.

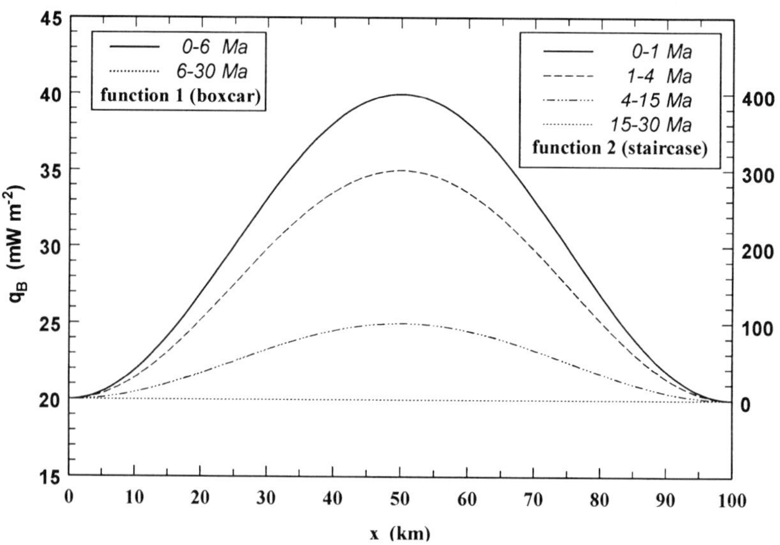

Fig. 4.4. Lateral variation of the boxcar and the staircase functions (Fig. 4.3) for q_B (heat flow density) and m_B (mass flow density) at the base of the model. The scale for mass flow density corresponds to a water content of 6 wt.% in the magma of Fig. 4.2. The legend box on the left shows the variation with time for function 1, the one on the right for function 2.

The boxcar function 1 is also used for specifying the mass flow density m_B at the base of the model, when required. Further details of the modeling are described by Clauser and Neugebauer (1991).

The model response is determined not only by the initial and boundary conditions, but also by the parameter distribution selected. For this reason, several conceptual models differing in their complexity were investigated to study the effects of anisotropy, fault zones, the sediment cover, and the mechanisms of free and forced convection. In all of the models the permeability was varied over three orders of magnitude in order to study the transition from a conductive thermal regime to an advection dominated one. All other rock properties were held constant, except for those that were varied according to the temperature and pressure conditions during the computations, as mentioned above.

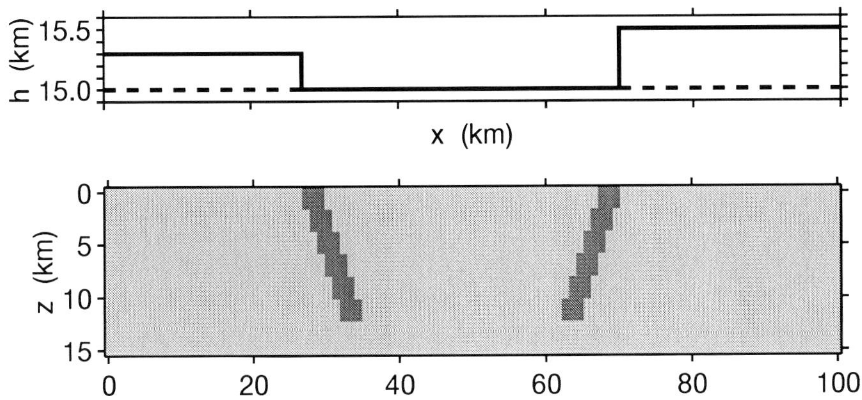

Fig. 4.5. Models studied: Model I with a constant hydraulic potential h at the top boundary (dashed line) and a homogeneous structure; Model II with an asymmetric upper boundary condition for the hydraulic potential (solid line) and fault zones added to the homogeneous structure.

The considerable influence of permeability on the flow field and the associated temperature distribution and total heat transport can now be illustrated using a homogeneous and isotropic model (model I). A second model (model II) that differs from the first one by two fault zones of increased permeability at the sides of the model and an asymmetric hydraulic head upper boundary condition illustrates the effect of the channeling of fluid flow through tectonic fracture zones (Fig. 4.5). It also permits differentiation between convection forced by hydraulic potential gradients and free convection resulting from differences in density. In all of the models the permeability was systematically varied over a range of three orders of magnitude (10^{-16} m^2 $\leq$ k $\leq$ 10^{-18} m^2). Permeability in the fault zones is increased a hundred times in respect to the surrounding rock. Porosity, rock compressibility, and heat production rate were held constant at $\Phi = 1\ \%$, $\alpha = 10^{-11}\ \mathrm{Pa}^{-1}$, and $H = 3\ \mu\mathrm{Wm}^{-3}$, respectively. For updating thermal conductivity with temperature a reference value of $\lambda(T=20\ °C) = 2.5\ \mathrm{W\ m^{-1}\ K^{-1}}$ was used for thermal conductivity at 20 °C.

The computation results for model I are shown in Plates 20 – 22. The temperature and flow fields for a permeability of $10^{-16}\ \mathrm{m}^2$ are shown in Plate 20. The thermal anomaly was simulated by heating function 2. After a simulation time of about 2 Ma, friction is overcome by the thermal buoyancy resulting from the heating, and strong free convection develops. Beginning at 3 Ma, warm water ascends to the surface, at first only directly above the center of the anomaly. At 15 Ma, the thermal anomaly has faded to the background level, and symmetric flow and temperature fields have developed. Owing to the inertia of the system, convection continues for a long time, with the convection cells becoming as far apart as possible.

If heating function 1 is used instead of heating function 2, a downwards flow and, thus, a temperature minimum develops at the center of the thermal anomaly after this anomaly has faded completely. Only after a long simulation period does the direction of flow reverse (Clauser and Neugebauer 1991). This is apparently due to a "cut-off effect", causing the solution to converge at first to a secondary minimum, owing to the special form of the heating function (rectangular pulse). This effect is avoided when the more realistic heating function 2 is used.

When the permeability is decreased by one order of magnitude, frictional forces predominate, and hence no free convection can develop (Plate 21). The flow velocities, however, are more than two orders of magnitude smaller, an effect of the nonlinearity caused by the temperature dependence of the water density and viscosity. After 30 Ma, twice the duration of the thermal disturbance, the anomaly has completely faded, mainly by diffusive temperature equilibration.

Magma contains a certain amount of bound H_2O. Depending on the type of magma, water contents of 1 – 6 wt.% are reported in the literature (Candela and Holland 1986). Sood (1981) gives 1 – 2.5 wt.% for early eruption phases, 0.2 – 0.7 wt.% for late eruption phases, and about 0.5 wt.% for basalt at depth. To estimate the maximum effect of the dehydration, maximum water contents of 1.5 and 6 wt.% were used in the calculations. For the given geometry of the intrusion (Fig. 4.2), this corresponds to 3×10^{13} kg and 12×10^{13} kg H_2O. Assuming steady-state conditions and the same lateral variation as used for heat flow density at the base of the model (function 1, Fig. 4.3), a mass flow density between 0 and a maximum of 100 or 400 $\mathrm{g\ a^{-1}\ m^{-2}}$, respectively, is obtained (Fig. 4.4). To isolate the thermal effect of dehydration from that of the thermal anomaly, (1) a permeability of $10^{-18}\ \mathrm{m}^2$ is used, in order to suppress any thermally induced free convection, and (2) the steady-state temperature distribution of the corresponding, purely conductive model is used as thermal boundary condition at the base. This guarantees that advection of heat can be caused only by the unidirectional forced convection system brought about by the release of water of crystallization. The effect caused by the release of 6 wt.% of water as the pluton cools is illustrated in Plate 22. This corresponds to the mass flow density across the base of the model shown in Figure 4.4. While the geometry of the resulting flow system is completely different from that of the previous models, its effect on the temperature field in this model cannot be recognized in Plate 22. A Péclet number analysis, however, yields a maximum value of 0.5. This means that advection contributes 1/3 of the total heat transport.

The objective of simulations using model II is to study the influence of faults at the lateral boundaries on the flow and temperature fields. The faults in this model increase the permeability of the rock by as much as two orders of magnitude. The asymmetrical boundary condition for hydraulic potential at the top of the model is also closer to the actual situation than model I. Because the undesired effect described above resulting from the use of heating function 1 does not occur in this model, heating function 1 was used for all the subsequent simulations. Because it delivers the same amount of heat into the system as heating function 2, but in a much shorter period of time, the model reacts to this input more strongly. This is clearly illustrated in Plate 23 for a permeability of 10^{-16} m^2. The ascending flow in the central part of the anomaly has reached its full development after 2 Ma and is more pronounced than in the corresponding example in Plate 20. The results reflect the asymmetry of the upper boundary condition for hydraulic potential. The channeling effect of the fault zones at the sides of the model can be distinctly seen. This is even more pronounced in the model with a permeability an order of magnitude smaller (Plate 24). In contrast to the homogeneous model I, the isotherms here indicate that convection is still thermally relevant. This is, however, limited to the area of the fault zones. Thus, this model shows a distinctly different behavior from that of the homogeneous model I.

The influence of free convection and the temperature dependence of the physical properties of water and rock can be seen in the difference between the results of two simulations with the same model: The one simulation takes the temperature dependence of the parameters into consideration, whereas the other uses constant coefficients. Comparison of the results shows primarily the influence of free convection on the simulations. This is illustrated in Plate 25 for model II with a permeability of 10^{-16} m^2, i.e., for the simulation in Plate 23. Especially at early simulation times, heat is transported predominantly by free convection, as evidenced by maximum differences of more than -100 K and +200 K between the temperature fields of the two simulations. But even long after the end of the thermal anomaly at 6 Ma, say at 30 and even 100 Ma, its contribution is still on the order of ± 50 K.

Permeabilities of 10^{-16} m^2 are certainly not realistic for a crustal sequence 15 km thick. Therefore, it is even more important to be able to distinguish whether circulation in the area of fault zones is caused by free or forced convection: A thermal signal from great depth can be advected to the surface only by free convection. Because free convection flow originates at depth, it does not have to descend first, like forced convection flow, in order to pick up and store heat at depth. Moreover, it can transport a signal sufficiently rapidly before it is too diffuse to recognize it.

This distinction cannot be made, however, using purely thermal data and methods. Additional information is necessary from other parameters that are sensitive to fluid flow and that can be unambiguously correlated with the lower crust or mantle. Isotopes, especially those of helium and argon, are useful for this purpose, as their origin can be attributed to distinct source regions, such as the atmosphere, the crust and the lower and upper mantle mantle (O'Nions and Oxburgh (1988), Matsuda (1994)). Griesshaber et al. (1992), and Darling et al. (1995) discuss this aspect using recent data as well as data from the literature for several Central European graben systems. A combination of thermal and isotope hydrology data and methods, therefore, appears to be especially suitable for identifying signals from thermal and tectonic events in the deep crust.

References

Arhavskaya, N. I., Galdin N. E., Karus, E. W., Kuznetsov, O. L., Lubimova, E. A., Milanovsky, S. Yu., Nartikoev, V. D., Semashko, S. A. and Smirnova, E. V., 1987. Geothermic investigations, in *The superdeep well of the Kola Peninsula,* pp. 387-393, edited by Kozlovsky, Ye. A., Springer Verlag, Berlin-Heidelberg (Germany).

Barker, C. E., 1989. Temperature and time in the thermal maturation of sedimentary organic matter, in *Thermal history of sedimentary basins,* pp. 73-98, edited by Naeser, N. D. and McCulloh, T. H., Springer Verlag, New York NY (USA).

Bayuk, E. I., Belikov, B. P., Vernik, L. I., Volarovitch, M. P., Kuznetsov, Yu. I., Kuzmenkova, G. E. and Pavlova, N. N., 1987. Rock density, porosity and permeability, in *The superdeep well of the Kola Peninsula,* pp. 332-338, edited by Kozlovsky, Ye. A., Springer Verlag, Berlin-Heidelberg (Germany).

Bear, J., 1972. *Dynamics of Fluids in Porous Media,* American Elsevier, New York NY (USA).

Beck, A. E., 1977. Climatically perturbed temperature gradients and their effects on regional and continental heat-flow means, *Tectonophysics, 41*, 17-39.

Beck, A. E., 1982. Precision logging of temperature gradients and the extraction of past climate, *Tectonophysics, 83,* 1-11.

Beck, A. E. and Judge, A. S., 1969. Analysis of heat flow data - I. Detailed observations in a single borehole, *Geophys. J. R. Astr. Soc., 18,* 145-158.

Beck, A. E., 1988. Methods for determining thermal conductivity and thermal diffusivity in *Handbook of Terrestrial Heat Flow Density Determination,* pp. 87-124, edited by Hänel, R., Rybach, L. and Stegena, L., Kluwer, Dordrecht (The Netherlands).

Beltrami, H., Jessop, A. M. and Mareschal, J.-C., 1992. Ground temperature histories in eastern and central Canada from geothermal measurements, *Global and Planetary Change, 6,* 167-184.

Beltrami, H. and Mareschal, J.-C., 1991. Recent warming in Eastern Canada: Evidence from geothermal measurements, *Geophys. Res. Lett., 18,* 605-608.

Beltrami, H. and Mareschal, J.-C., 1992. Ground temperature histories for central and eastern Canada from geothermal measurements: Little Ice Age signature, *Geophys. Res. Lett., 19,* 689-692.

Benfield, A. E., 1939. Terrestrial heat-flow in Great Britain, *Proc. R. Soc. London A, 173,* 428-450.

Berckhemer, H., Bücker, C., Huenges, E., Lippmann, E., Rauen, A., Streit, K. M., Wienand, J. and Soffel, H. C., 1997. Petrophysical properties of the 9 km deep crustal section at KTB, *J. Geophys. Res, 102(B8),* 18337-18361.

Berger, A., Imbrie, J., Kukla, G. and Saltzman, B. (eds), 1984. *Milankovitch and Climate, Parts. I & II,* Reidel, Dordrecht (The Netherlands).

Bethke, C. M., 1985. A numerical model of compaction-driven groundwater flow and heat transfer and its application to the paleohydrology of intracratonic sedimentary basins, *J. Geophys. Res, 90(B8),* 6817-6828.

Birch, F., 1948. The effects of pleistocene climatic variations upon geothermal gradients, *Am. J. Sci., 246*, 729-760.

Birch, F., Roy, R. F. and Decker, E. R., 1968. Heat flow and thermal history in New England and New York, in *Studies of Appalachian Geology*, pp. 437-451, edited by Zen, E-an, White, W. S., Hadley, J. B. and Thompson Jr., J. B., Interscience, New York NY (USA).

Borevsky, L. V., Vartanyan, G. S. and Kulikov, T. B., 1987. Hydrogeological essay, in *The superdeep well of the Kola Peninsula,* pp. 271-287, edited by Kozlovsky, Ye. A., Springer Verlag, Berlin-Heidelberg (Germany).

Bredehoeft, J. D. and Papadopulos, I. S., 1965. Rates of vertical groundwater movement estimated from the earth's thermal profile, *Water Resour. Res., 1(2),* 325-328.

Bullard, E. C., 1939. Heat flow in South Africa, *Proc. Roy. Soc. London A, 173,* 474-502.

Buntebarth, G., 1984. *Geothermics,* Springer Verlag, Berlin-Heidelberg (Germany).

Buntebarth, G., 1991. Thermal properties of KTB-Oberpfalz VB core samples at elevated temperature and pressure, *Scientific Drilling, 2,* 73-80.

Buntebarth, G. and Stegena, L. (eds), 1986. *Paleogeothermics,* Lecture notes in earth sciences, 5, Springer Verlag, Berlin-Heidelberg (Germany).

Burkhardt, H., Haack, U., Hahn, A., Honarmand, H., Jäger, K., Stiefel, A., Wägerle, P. and Wilhelm, H., 1989. Geothermal investigations at the KTB Locations Oberpfalz and Schwarzwald, in *The German Continental Deep Drilling Program (KTB): Site Selection Studies in the Oberpfalz and Schwarzwald,* pp. 433-480, edited by Emmermann, R. and Wohlenberg, J., Springer Verlag, Berlin-Heiodelberg (Germany).

Burrus, J. and Audebert, F., 1990. Thermal compaction processes in a young rifted basin containing evaporites: Gulf of Lions, France, *AAPG Bull., 74(9),* 1420-1440.

Burruss, R. C., 1989. Paleotemperatures from fluid inclusions: advances in theory and technique, in *Thermal history of sedimentary basins,* pp. 119-131, edited by Naeser, N. D. and McCulloh, T. H., Springer Verlag, New York NY (USA).

Candela, P. A. and Holland, H. D., 1986. A mass transfer model for copper and molybdenum in magmatic hydrothermal systems: the origin of porphyry-type ore deposits, *Econ. Geology, 81(1)*, 1-19.

Carslaw, H. S. and Jaeger, J. C., 1959. *Conduction of Heat in Solids, 2nd ed.,* Oxford University Press, Oxford (United Kingdom).

Čermák, V., 1971. Underground temperature and inferred climatic temperature of the past millennium, *Palaeogeogr., Palaeoclimatol., Palaeoecol., 10,* 1-19.

Čermák, V., 1975. Combined heat flow and heat generation measurements in the Bohemian Massif, *Geothermics*, *4(1-4)*, 19-26.

Čermák, V., 1981. Heat flow investigations in Czechoslovakia, in *Geophysical Syntheses in Czechoslovakia,* pp. 427-439, edited by Zátopek, A., Petr, V. and Vaňk, J., Veda - publishing house of the Slovak Academy of Sciences.

Čermák, V., Bodri, L. and Šafanda, J., 1992. Recent climate change recorded in the underground: evidence from Cuba, *Global and Planetary Change, 6,* 219-224.

Chapman, D. S., Howell, J. and Sass, J. H., 1984. A note on drillhole depths required for reliable heat flow determinations, in *Terrestrial heat flow studies and the structure of the lithosphere*, edited by Čermák, V., Rybach, L. and Chapman, D. S., *Tectonophysics, 103(1-4),* 11-18.

Chapman, D. S. and Rybach, L., 1985. Heat flow anomalies and their interpretation, in *Heat flow and geothermal processes,* edited by Rybach, L., *J. Geodynamics, 4(1-4),* 3-37.

Chen, Y. and Beck, A. E., 1991. Application of the boundary element method to a terrestrial heat flow problem, *Geophys. J. Int., 107(1),* 25-35.

Clauser, C., 1984. A climatic correction on temperature gradients using surface temperature series of various periods, in *Terrestrial Heat Flow and the Structure of the Lithosphere*, edited by Čermák, V., Rybach, L. and Chapman D. S., *Tectonophysics, 103(1-4)*, 33-46.

Clauser, C., 1988a. Untersuchungen zur Trennung der konduktiven und konvektiven Anteile im Wärmetransport in einem Sedimentbecken am Beispiel des Oberrheintalgrabens, Doctoral Dissertation, Tech. Univ. Berlin (Germany), *Fortschritt-Berichte VDI, Reihe 19, Nr. 28*, VDI Verlag, Düsseldorf (Germany).

Clauser, C., 1988b. Opacity - the Concept of Radiative Thermal Conductivity, in *Handbook of Terrestrial Heat Flow Density Determination,* pp. 143-165, edited by Hänel, R., Rybach L. and Stegena L., Kluwer Academic Publishers, Dordrecht (The Netherlands).

Clauser, C., 1992. Permeability of crystalline rocks, *EOS, Trans. Am. Geophys. Union, 73,* 233,237.

Clauser, C., Giese, P., Huenges, E., Kohl, T., Lehmann, H., Rybach, L., Šafanda, J., Wilhelm, H., Windloff, K., and Zoth, G., 1997. The thermal regime of the crystalline continental crust: Implications from the KTB, *J. Geophys. Res., 102(B8),* 18417-18441.

Clauser, C. and Huenges, E., 1993. KTB thermal regime and heat transport mechanisms - current knowledge, *Scientific Drilling, 3(6),* 271-281.

Clauser, C. and Huenges, E., 1995. Thermal conductivity of Rocks and Minerals, in *Handbook of Physical Constants, Vol. 3: Rock Physics and Phase Relations*, edited by T. J. Ahrens, Amer. Geophys. Union, Washington D.C. (USA).

Clauser, C. and Kiesner, S., 1987. A Conservative, unconditionally stable, second-order three-point differencing scheme for the diffusion-convection Equation. *Geophys. J. R. astr. Soc., 91*, 557-568.

Clauser, C. and Mareschal, J.-C., 1995. Ground temperature history in Central Europe from borehole temperature data, *Geophys. J. Int., 121(3),* 805-817.

Clauser, C. and Neugebauer, H. J., 1991. Thermisch relevante Tiefenwasserzirkulation in der Oberkruste unter dem Oberrheingraben? Eingrenzungen mit Hilfe hydrothermischer Modellrechnungen. *Geoogisches. Jahrbuch, Reihe E, Vol. 48, pp.* 185-217, Hannover (Germany).

Clauser, C. and Villinger, H., 1990. Analysis of conductive and convective heat transfer in a sedimentary basin, demonstrated for the Rheingraben. *Geophys. J. Int., 100(3),* 393-414.

Creaney, S., 1989. Reaction of organic material to progressive geological heating, in *Thermal history of sedimentary basins,* pp. 37-52, edited by Naeser, N. D. and McCulloh, T. H., Springer Verlag, New York NY (USA).

Curiale, J. A., Larter, S. R., Sweeney, R. E. and Bromley, B. W., 1989. Molecular thermal maturity indicators in oil and gas source rocks, in *Thermal history of sedimentary basins,* pp. 53-72, edited by Naeser, N. D. and McCulloh, T. H., Springer Verlag, New York NY (USA).

Darcy, H., 1856. *Les fontaines publiques de la ville de Dijon,* Dalmont, Paris (France).

de Marsily, G., 1986, Quantitative Hydrogeology, Academic Press, Orlando FLA (USA).

Darling G., Griesshaber E., Andrews N., Armannsson H. and O'Nions R. K., 1995. The origin of hydrothermal and other gases in the Kenya Rift Valley, *Geochim. Cosmochim. Acta, 59,* 2501-2512.

Delisle, G., 1991. Causes and consequences of the advance of a continental ice sheet from Scandinavia to Northern Germany, *Geol. Jb., Ser. A, 127,* 507-517.

Düppenbecker, S. J. and Welte, D. H., 1992. Petroleum expulsion from source rocks - insights from geology, geochemistry and cumputerized numerical modelling, in *Proc. 13. World*

Petroleum Congress Buenos Aires 1991, Vol. 2, pp. 165-177, John Wiley & Sons, Chichester (United Kingdom).

Emmermann, R. and Wohlenberg J. (eds) 1989. *The German Continental Deep Drilling Program (KTB): Site-selection Studies in the Oberpfalz and Schwarzwald,* Springer Verlag, Berlin-Heidelberg (Germany).

England, P. C. and Thompson, A. B., 1984. Pressure-temperature-time paths of regional metamorphism, Part 1: Heat Transfer during the evolution of regions of thickened continental crust, *J. Petrol., 25(4),* 894-928.

England, P. C. and Thompson, A. B., 1986. Some thermal and tectonic models for crustal melting in continental collision zones, in *Collision Tectonics,* edited by Coward, M. P. and Ries, A. C., *Geol. Soc. Lond. Spec. Publ., 19,* 83-94.

Epstein, S., Buchsbaum, R., Lowenstam, H. A. and Urey, H. C., 1953. Revised carbonate-water isotopic temperature scale, *Geol. Soc Am. Bull, 64,* 1315-1325.

Espitalié, J., Laporte, J. L., Madec, M., Marquis, F., Leplat, P., Paulet, J. and Boutefeu, A., 1977. Méthode rapide de charactérisation des roches mères, de leur potentiel pétrolier et de leur degré d'évolution, *Rev. Inst. Fr. Pét., 32,* 23-42.

Faegri, K. and Iversen, J., 1989. *Textbook of Pollen Analysis, 4th ed.,* revised by Faegri, K., Kaland, P. E. and Krzywinski, K., John Wiley & Sons, Chichester (United KIngdom).

Faure, G., 1986. *Principles of isotope geology, 2nd ed.,* John Wiley & Sons, New York NY (USA).

Forster, C. and Smith, L., 1989. The influence of groundwater flow on thermal regimes in mountainous terrain: a model study, *J. Geophys. Res., 94(B7),* 9439-9451.

Forster, C. and Smith, L., 1990. Fluid flow in tectonic regimes, in Fluids in tectonically *active Regimes of the Continental Crust,* pp. 1-47, Short Course Handbook, Vol. 18, edited by Nesbitt, B. E., Mineralogical Association of Canada, Nepean, Ontario (Canada).

Fournier, R. O., 1981. Application of water chemistry to geothermal exploration and reservoir engineering, in *Geothermal Systems: Principles and Case Histories,* pp. 109-143, edited by Rybach, L. and Muffler, L. J. P., John Wiley and Sons, Chichester (United Kingdom).

Fournier, R. O. and Truesdell, A. H., 1973. An empirical Na-K-Ca geothermometer for natural waters, *Geochim. Cosmochim. Acta, 37,* 1255-1275.

Galdin, N. Ye., Nartikoyev, V. D., Semashko, S. V., Popov, Yu. A., Korostelev, V. M. and Berezin, V. V., 1986. Thermal conductivities of dry and water-saturated low-porosity crystalline rocks of the Archaean Kola Series, *Int. Geol. Rev., 28,* 858-865.

Galdin, N. E., Lubimova, E. A., Nartikoyev, V. D., Popov, Y. A., Semashko, S. A., Skornyakov, V. M. and Smirnova, E. V., 1987. Rock thermal properties, in *The superdeep well of the Kola Peninsula,* pp. 383-386, edited by Kozlovsky, Ye. A., Springer Verlag, Berlin-Heidelberg (Germany).

Garven, G. and Freeze, A., 1984a. Theoretical analysis of the role of groundwater flow in the genesis of stratabound ore deposits - 1. Mathematical and numerical model, *Am. J. Sci., 284,* 1085-1124.

Garven, G. and Freeze, A., 1984b. Theoretical analysis of the role of groundwater flow in the genesis of stratabound ore deposits - 2. Quantitative results, *Am. J. Sci., 284,* 1125-1174.

Geyh, M. A. and Schleicher, H., 1990. *Absolute age determination,* Springer Verlag, Berlin-Heidelberg (Germany).

Green, D. H., 1970. A review of experimental evidence on the origin of basaltic and nephelinitic magmas, *Phys. Earth Planet. Interiors, 3,* 221-235.

Griesshaber E., O'Nions R. K. and Oxburgh E. R., 1992. Helium and carbon isotope systematics in crustal fluids from the Eifel, Rhine Graben and Black Forest, *Chem. Geol., 99,* 213-235.

Gudden, H. and Schmidt, H., 1985. Die Forschungsbohrung Obernsees (westlich von Bayreuth): Konzeption, Durchführung und Untersuchung der Metallführung, *Geologica Bavarica, 88,* 5-21, München (Germany).

Haack, U., 1977. The closing temperature for fission track retention in minerals, *Am. J. Sci., 277,* 459-464.

Hänel, R. and Mongelli, F., 1988. Thermal exploration methods, in *Handbook of Terrestrial Heat Flow Density Determination,* pp. 353-389, edited by Hänel, R., Rybach, L. and Stegena, L., Kluwer, Dordrecht (The Netherlands).

Hänel, R., and Staroste, E. (eds), 1988. *Atlas of geothermal resources in the European Community, Austria, and Switzerland*, Verlag Th. Schäfer, Hannover (Germany).

Hansen, J. and Lebedeff, S., 1987. Global trends of measured air-surface temperature, *J. Geophys. Res., 92*, 13345-13372.

Hawkesworth, C. and van Calsteren, P., 1992. Geological time, in *Understanding the earth,* pp. 132-144, edited by Brown, G., Hawkesworth, C. and Wilson, C., Cambridge University Press, Cambridge (United Kingdom).

Hirschmann, G., 1993. Zur Geologie der KTB-Lokation Oberpfalz, *Z. geol. Wiss., 21(1/2),* 105-116.

Hollister, L. S. and Crawford, M. L. (eds), 1981. *Fluid inclusions: Applications to petrology,* Short Course Handbook, Vol. 6, Mineralogical Association of Canada, Toronto (Canada).

Hood, A., Gutjahr, C. C. M. and Heacock, R. L., 1975. Organic metamorphism and the generation of petroleum, *AAPG Bull., 59,* 986-996.

Hotchkiss, W. O. and Ingersoll, L. R., 1934. Post glacial time calculations from recent measurements in the Calumet Copper mine, *J. Geol., 42,* 113-142.

Huenges, E., Burkhardt, H. and Erbaş, K., 1990. Thermal conductivity profile of the KTB pilot corehole, *Scientific Drilling, 1,* 224-230.

Huenges, E, Erzinger, J., Kück, J., Engeser, B. and Kessels, W., 1997. The permeable crust: Geohydraulic properties down to 9101 m depth, *J. Geophys. Res., 102(B8),* 18255-18265.

Imbrie, J., Hays, J. D., Martinson, D. G., McIntyre, A., Mix, A. C., Morley, J. J., Pisias, N. G., Prell, W. L. and Shackleton, N. J., 1984. The orbital theory of Pleistocene climate: support from a revised chronology of the marine $\delta^{18}O$ record, in *Milankovitch and Climate, Part I,* pp. 269-305, edited by Berger, A., Imbrie, J., Kukla, G. and Saltzman, B., Reidel, Dordrecht (The Netherlands).

Jackson, D. D., 1972. Interpretation of inaccurate, insufficient, and inconsistent data, *Geophys. J. R. Astr. Soc., 28,* 97-110.

Jessop, A., 1971. The distribution of glacial perturbation of heat-flow in Canada, *Can. J. Earth Sci., 5,* 61-68.

Jobmann, M. and Clauser, C., 1994. Heat advection versus conduction at the KTB: possible reasons for vertical variations in heat flow density, *Geophys. J. Int., 119(1),* 44-68.

Kappelmeyer, O. and Hänel, R., 1974. *Geothermics - with Special Reference to Application,* Verlag Gebrüder Bornträger, Berlin (Germany).

Kazansky, V. I., Kremenetsky, A. A., Lanev, V. S., Nalivkina, E. B. and Ovchinnikov, L. N., 1987. Evolution of the continental crust in the Precambrian, in *The superdeep well of the Kola Peninsula,* pp. 288-312, edited by Kozlovsky, Ye. A., Springer Verlag, Berlin-Heidelberg (Germany).

Kharaka, Y. K. and Mariner, R. H., 1989. Chemical geothermometers and their application to formation waters from sedimentary basins, in *Thermal history of sedimentary basins,* pp. 99-117, edited by Naeser, N. D. and McCulloh, T. H., Springer Verlag, New York NY (USA).

Kohl, T., Hopkirk, R. J., Evans, K. F. and Rybach, L., 1993. FRACTure - a new tool to simulate coupled processes in geosciences, in *Finite Elements in Fluids,* pp. 949-958, edited by Morgan, K., Oñate, E., Periaux, J., Peraire, J. and Zienkiewicz, O. C., Pineridge Press, Mumbles, Swansea (United Kingdom).

Kohl, T. and Rybach, L., 1995. Thermal and hydraulic aspects of the KTB drill site, *Geophys. J. Int.,124(3),* 756-772.

Kozlovsky, Ye. A. (ed), 1987. *The superdeep well of the Kola Peninsula,* Springer Verlag, Berlin-Heidelberg (Germany).

Kremenetsky, A. A. and Krivtsov, A. I. (eds), 1991. *Models and cross-sections of the Earth's crust based on superdeep drilling data of the USSR,* Imgre, Moscow (Russia).

Kremenetsky, A. A. and Ovchinnikov, L. N., 1986a. The Precambrian continental crust: Its structure, composition and evolution as revealed by deep drilling in the USSR, *Precambrian Res., 33,* 11-43.

Kremenetsky, A. A. and Ovchinnikov, L. N., 1986b. *Geohimiya glubinnyh porod,* Nauka, Moscow (Russia).

Kremenetsky, A. A., Milanovsky, S. Yu. and Ovchinnikov, L. N., 1989. A heat generation model for continental crust based on deep drilling in the Baltic Shield, *Tectonophysics, 159,* 231-246.

Kukkonen, I. T. and Clauser, C., 1994. Simulation of heat transfer at the Kola deep-hole site - implications for advection, heat refraction and palaeoclimatic effects, *Geophys. J. Int., 116(2):* 409-420.

Kukowski, N. and Neugebauer, H. J., 1990. On the ascent and emplacement of granitoid magma bodies - dynamic-thermal numerical models, *Geol. Rundschau, 79(2),* 227-239.

Lachenbruch, A. H., 1968. Preliminary geothermal model of the Sierra Nevada, *J. Geophys. Res., 73(22),* 6977-6989.

Lachenbruch, A. H. and Marshall, B. V., 1986. Changing climate: geothermal evidence from permafrost in the Alaskan arctic, *Science, 234,* 689-696.

Lanczos, C., 1961. *Linear Differential Operators,* D. van Nostrand, New York NY (USA).

Lane, A. C., 1923. Geotherms from the Lake Superior copper country, *Bull. Geol. Soc. Am., 34,* 703-720.

Lanev, V. S., Nalivkina, E. B., Vakhrusheva, V. V., Golenkina, E. A., Rusanov, M. S., Smirnov, Y. P., Suslova, S. N., Duk, G. G., Koltsova, T. V., Maslennikov, V. A., Timofeev, B. V. and Zaslavsky, V. G., 1987. Geological Section of the well, in *The superdeep well of the Kola Peninsula,* pp. 40-73, edited by Kozlovsky, Ye. A., Springer Verlag, Berlin-Heidelberg (Germany).

Lehmann, H., Wang, K., and Clauser, C., 1998. Parameter identification and uncertainty analysis for heat transfer at the KTB drill site using a 2-D inverse method, *Tectonophysics, 291,* 179-194.

Leischner, K., 1994. Kalibration simulierter Temperaturgeschichten von Sedimentgesteinen. Doctoral Dissertation, Ruhr-Universität Bochum (Germany), *Bericht Jül-2909,* Institut für Chemie und Dynamik der Lithosphäre 4: Erdöl und Organische Geochemie, Forschungszentrum Jülich (Germany).

Leischner, K., Welte, D. H. and Littke, R., 1993. Fluid inclusions and organic maturity parameters as calibration tools in basin modelling, in *Basin Modelling: Advances and Applications,* pp. 161-172, edited by Doré, A. G., Augustson, J. H., Hermanrud, C., Steward, D. J. and Sylta, Ø., Elsevier, Amsterdam (The Netherlands).

Lewis, T. J. (ed), 1992.. Climatic changes inferred from underground temperatures, *Global and Planetary Change, 6 (Special Volume).*

Mansure, A. J. and Reiter, M., 1979. A vertical groundwater correction for heat flow, *J. Geophys. Res., 84(B7),* 3490-3496.

Mareschal, J.-C. and Beltrami, H., 1992. Evidence for recent warming from perturbed geothermal gradients: examples from eastern Canada, *Clim. Dyn., 6*, 135-143.

Mareschal, J.-C. and Vasseur, G., 1992. Ground temperature history from two deep boreholes in central France, *Global and Planetary Change, 6,* 185-192.

Martinson, D. G., Pisias, N. G., Hays, J. D., Imbrie, J., Moore, T. C. and Shackleton, N. J., 1987. Age dating and the ohe orbital theory of the ice ages: development of a high-resolution 0 to 300,000-year chronostratigraphy, *Quarternary Res., 27,* 1-29.

Matsuda J., 1994. *Noble gas geochemistry and cosmochemistry,* Terra, Tokyo (Japan).

Menke, W., 1989. *Geophysical Data Analysis: Discrete Inverse Theory, 2nd ed.,* International Geophysical Series, 45, Academic Press, San Diego CA (USA).

Möller, P., Weise, S. M., Althaus, E., Bach, W., Behr, H. J., Borchardt, R., Bräuer, K., Drescher, J., Erzinger, J., Faber, E., Hansen, B. T., Horn, E. E., Huenges, E., Kämpf, H., Kessels, W., Kirsten, T., Landwehr, D., Lodemann, M., Machon, M., Pekdeker, A., Pielow, H.-U., Reutel, C., Simon, K., Walther, J., Weinlich, F. H. and Zimmer, M., 1997. Paleofluids and recent fluids in the uppercontinental crust: Results from the German Continental Deep Drilling Program (KTB), *J. Geophys. Res., 102(B8),* 18233-18254.

Moore, P. D., Webb, J. A. and Collinson, M. E., 1991. *Pollen Analysis, 2nd ed.,* Blackwell Scientific Publications, Oxford (United Kingdom).

Naeser, C. W., 1979. Fission-track dating and geologic annealing of fission tracks, in *Lectures in Isotope Geology,* pp. 154-169, edited by Jäger, E. and Hunziker, J. C., Springer Verlag, New York NY (USA).

Naeser, N. D., Naeser, C. W. and McCulloh, T. H., 1989. The application of fission-track dating to the depositional and thermal history of rocks in sedimentary basins, in *Thermal history of sedimentary basins,* pp. 157-180, edited by Naeser, N. D. and McCulloh, T. H., Springer Verlag, New York NY (USA).

Nalivkina, E. B., Rusanov, M. S. and Suslova, S. N., 1987. Geology of the north-eastern part of the Baltic Shield, in *The superdeep well of the Kola Peninsula,* pp. 17-39, edited by Kozlovsky, Ye. A., Springer Verlag, Berlin-Heidelberg (Germany).

Negendank, J. F. W. and Zolitschka, B. (eds), 1993. *Paleolimnology of European maar lakes,* Lecture notes in earth sciences, 49, Springer Verlag, Berlin-Heidelberg (Germany).

Nielsen, S. B. and Beck, A. E., 1989. Heat flow density values and paleoclimate determined from stochastic inversion of four temperature-depth profiles from the Superior Province of the Canadian shield, *Tectonophysics, 164*, 345-359.

O'Neil, J. R., 1979. Stable isotope geochemistry of rocks and minerals, in *Lectures in Isotope Geology,* pp. 235-263, edited by Jäger, E. and Hunziker, J. C., Springer Verlag, New York NY (USA).

O'Nions R. K. and Oxburgh E. R., 1988. Helium, volatile fluxes and the development of the continental crust, *Earth Planet. Sci. Lett., 90*, 331-347.

O'Reilly, S. Y. and Griffin, W. L., 1990. Geophysical and petrologic properties of the crust/mantle boundary region, eastern Australia: relevance to the Eromanga-Brisbane transect, *BMR Bull., 232,* 203-212, Australian Bureau of Mineral Resources.

Pearson, N. J. and O'Reilly, S. Y., 1991. Thermobarometry and P-T-t paths: the granulite to eclogite transition in lower crustal xenoliths from eastern Australia, *J. Metamorphic Geol., 9,* 349-359.

Pearson, N. J., O'Reilly, S. Y. and Griffin, W. L., 1991. Heterogeneity in the thermal state of the lower crust and upper mantle beneath eastern Australia, *Exploration Geophysics, 22,* 295-298.

Person, M. and Garven, G., 1992. Hydrologic constraints on petroleum generation within continental rift basins: theory and application to the Rhine Graben, *AAPG Bull., 76(4),* 468-488.

Person, M. and Garven, G., 1994. A sensitivity study of the driving forces on fluid flow during continental-rift basin evolution, *Geol. Soc. Am. Bull., 106,* 461-475.

Poley, J. P. and van Steveninck, J., 1970. Delineation of shallow salt domes and surface faults by temperature measurements at a depth of approximately 2 metres, *Geophys. Prosp., 18,* 666-700.

Pollack, H. N., Hurter, S. J. and Johnson, J. R., 1993. Heat flow from the earth's interior: analysis of the global data set, *Rev. Geophys., 31(3),* 267-280.

Press, W. H., Teukolsky, S. A., Vetterling, W. T. and Flannery, B. P., 1992. *Numerical Recipes in FORTRAN, 2nd ed.*, Cambridge University Press, Cambridge (United Kingdom).

Pribnow, D., 1994. Ein Vergleich von Bestimmungsmethoden der Wärmeleitfähigkeit unter Berücksichtigung von Gesteinsgefügen und Anisotropie, Doctoral Dissertation, Tech. Univ. Berlin (Germany), *Fortschritt-Berichte VDI, Reihe 19, Nr. 75*, VDI Verlag, Düsseldorf (Germany).

Pytte, A. M. and Reynolds, R. C., 1989. The thermal transformation of Smectite to Illite, in *Thermal history of sedimentary basins,* pp. 133-140, edited by Naeser, N. D. and McCulloh, T. H., Springer Verlag, New York NY (USA).

Roedder, E., 1984. *Fluid inclusions,* Reviews in Mineralogy, Vol. 12, Mineralogical Society of America, Washington D.C. (USA).

Roy, R. F., Blackwell, D. D. and Birch, F., 1968. Heat generation of plutonic rocks and continental heat flow provinces, *Earth Planet. Sci. Lett., 5,* 1-12.

Rudnick, R. L., 1992. Xenoliths - samples of the lower continental crust, in *Continental Lower Crust, Developments in Geotectonics, 23,* pp. 269-316, edited by Fountain, D. M., Arculus, A. and Kay, R. W., Elsevier, Amsterdam (The Netherlands).

Šafanda, J. and Kubik, J., 1992. Evidence from ground temperature changes from two boreholes in the Bohemian Massif, *Global and Planetary Change, 6,* 199-208.

Sass, J. H. and Lachenbruch, A. H., 1982. Preliminary interpretation of thermal data from the Nevada test site, *Open-File Report 82-973,* U. S. Geological Survey, Menlo Park CA (USA).

Schmoll, J., Bittner, R., Dürbaum, H.-J., Heinrichs, T., Meißner, R., Reichert, C., Rühl, T. and Wiederhold, H., 1989. Oberpfalz Deep Seismic Reflection Survey and Velocity Studies, in *The German Continental Deep Drilling Program (KTB): Site-Selection Studies in the Oberpfalz and Schwarzwald,* pp. 99-149, edited by Emmermann, R. and Wohlenberg, J., Springer Verlag, Berlin-Heidelberg (Germany) .

Schönwiese, C.-D., Rapp, J., Fuchs, T. and Denhard, M., 1994. Observed climate trends in Europe 1891-1990, *Meteorol. Zeitschrift, N. F. 3,* 22-28.

Schweingruber, F. H., 1988. *Tree Rings - Basics and Applications of Dendrochronology,* Reidel Publishing Company, Dordrecht (The Netherlands).

Sebagenzi, M. N., Vasseur, G. and Louis, P., 1992. Recent warming in southeastern Zaire (Central Africa) inferred from disturbed geothermal gradients, *Global and Planetary Change, 6,* 209-218.

Shen, P. Y. and Beck, A. E., 1983. Determination of ground temperature history from borehole temperature gradients, *J. Geophys. Res., 88,* 7385-7493.

Shen, P. Y. and Beck, A. E., 1991. Least squares inversion of borehole temperature measurements in functional space, *J. Geophys. Res., 96,* 19965-19979.

Smith, L. and Chapman, D. S., 1983. On the thermal effects of groundwater flow - 1. regional flow systems, *J. Geophys. Res., 88(B1),* 593-608.

Sood, M. K., 1981. *Modern Igneous Petrology,* John Wiley & Sons, New York NY (USA).

Stegena, L., 1988. Palaeogeothermics, in *Handbook of Terrestrial Heat Flow Density Determination,* pp. 391-419, edited by Hänel, R., Rybach L. and Stegena L., Kluwer Academic Publishers, Dordrecht (The Netherlands).

Stein, C. A., 1995. Heat flow of the earth, in *Global Earth Physics - a handbook of physical constants,* pp. 144-158, edited by Ahrens, T. J., American Geophysical Union, Washington D.C. (USA).

Sweeney, J. J. and Burnham, A. K., 1990. Evaluation of a simple model of vitrinite reflectance based on chemical kinetics, *AAPG Bull., 74(10),* 1559-1570.

Teichmüller, M. and Durand, B., 1983. Fluorescence microscopical rank studies on liptinites and vitrinites in peat and coals, and comparison with the results of Rock-Eval pyrolysis, *Int. J. Coal Geol., 2,* 197-230.

Thompson, A. B., 1992. Metamorphism and fluids, in *Understanding the earth,* pp. 222-248, edited by Brown, G., Hawkesworth, C. and Wilson, C., Cambridge University Press, Cambridge (United Kingdom).

Thompson, A. B. and England, P. C., 1984. Pressure-temperature-time paths of regional metamorphism, Part 2: Their inference and interpretation using mineral assemblages in metamorphic rocks, *J. Petrol., 25(4),* 929-955.

Tissot, B. P., Pelet, R. and Ungerer, P., 1987. Thermal history of sedimentary basins and kinetics of oil and gas generation, *AAPG Bull., 71(12),* 1445-1466.

Tissot, B. P. and Welte, D. H., 1984. *Petroleum occurrence and formation, 2nd ed.,* Springer Verlag, Berlin-Heidelberg (Germany).

Truesdell, A. H., 1976. Summary of section III: geochemical techniques in exploration, *Proc. Second United Nations Symposium on the Development and Use of geothermal Resources, San Francisco, 20-29 may 1975, Vol. 1,* pp. liii-lxxix, U.S. Government Printing Office, Washington, D.C. (USA).

Tuttle, O. F. and Bowen, N. L., 1958. Origin of granite in the light of experimental studies in the system $NaAlSi_3O_8$ - $KAlSi_3O_8$ - SiO_2 - H_2O, *Geol. Soc. Amer. Mem., 74,* 1-153.

Ungerer, P., Bessis, F., Chenet, P. Y., Durand, B., Nogaret, E., Chiarelli, A., Oudin, J. L. and Perrin, J. F., 1984. Geological and geochemical models in oil exploration; principles and practical examples, in *AAPG Memoir 35: Petroleum Geochemistry and Basin Evolution,* pp. 53-77, edited by Demaison, G. and Morris, R. J., Am. Ass. Petrol. Geol., Tulsa, OK (USA).

US Air Force, 1971. *Tactical pilotage chart C-2B, Finland, Norway, USSR, 1:500 000.* Aeronautical Information Center, U.S. Air Force, St. Louis, Missouri, 63118.

Vasseur, G., Bernard, P., Van de Meulebrouck, J., Kast, Y. and Jolivet, J., 1983. Holocene paleotemperatures deduced from geothermal measurements, *Palaeogeogr., Palaeoclimatol., Palaeoecol., 43,* 237-259.

Wagner, G. A. and Van den haute, P., 1992. *Fission-track dating,* Ferdinand Enke Verlag, Stuttgart (Germany).

Wang, K., 1992. Estimation of ground temperature history from borehole temperature data, *J. Geophys. Res., 97,* 2095-2106.

Wang, K. and Beck, A. E., 1989. An inverse approach to heat flow study in hydrologically active areas, *Geophys. J. Int., 98,* 69-84.

Wang, K. and Lewis, T. J., 1992. Geothermal evidence from Canada for a cold period before recent climatic warming, *Science, 256,* 1003-1005.

Waples, D. W., 1980. Time and temperature in petroleum formation: application of Lopatin's method to Petroleum exploration, *AAPG Bull., 64(6),* 916-926.

Welte, D. H., 1982. Neue Wege in der Kohlenwasserstoffexploration, *Erdöl und Kohle - Ergas - Petrochemie, 35(11),* 503-508.

Welte, D. H. and Yükler, M. A., 1981. Petroleum origin and accumulation in basin evolution - a quantitative model, *AAPG Bull., 65(8),* 1387-1396.

Willett, S. D. and Chapman, D. S., 1987. Analysis of temperatures and thermal processes in the Uinta Basin, in *Can. Soc. Petrol. Geol. Memoir 12: Sedimentary Basins and Basin-Forming Mechanisms,* pp. 447-461, edited by Beaumont, C. and Tankard, A. J., Can. Soc. Petrol. Geol.

Wood, D. A., 1988. Relationships between thermal maturity indices calculated using Arrhenius equation and Lopatin Method: implications for petroleum exploration, *AAPG Bull., 72(2),* 115-134.

Wyllie, P. J., 1988. Magma genesis, plate tectonics, and chemical differentiation of the earth, *Rev. Geophys., 26,* 370-404.

Wyllie, P. J., 1992. Experimental petrology: earth materials science, in *Understanding the earth,* pp. 67-87, edited by Brown, G., Hawkesworth, C. and Wilson, C., Cambridge University Press, Cambridge (United Kingdom).

Yalçin, M. N. and Welte, D. H., 1988. The thermal evolution of sedimentary basins and significance for hydrocarbon generation, *Turk. Ass. Petrol. Geol. Bull, 1(1),* 12-26.

Zolitschka, B., 1990. Spätquartäre jahreszeitlich geschichtete Seesedimente ausgewählter Eifelmaare, Doctoral Dissertation, University of Trier (Germany), *Documenta naturae, 60,* Verlag documenta naturae, München (Germany).

Zoth, G. and Hänel, R., 1988. Appendix, in *Handbook of Terrestrial Heat Flow Density Determination,* pp. 447-468, edited by Hänel, R., Rybach, L. and Stegena, L., Kluwer, Dordrecht (The Netherlands).

Appendix

Equations for Fluid Flow and Heat Transport in Porous Media

The empirically derived Darcy equation (Darcy 1856) describes the linear relationship between the specific discharge **v** (called the Darcy velocity) and the pressure gradient ∇P over a wide range of pressure gradients:

$$\mathbf{v} = \frac{-\mathbf{k}}{\mu}\left(\nabla P + \rho_f g \nabla z\right), \tag{24}$$

where ρ_f and μ are the density and dynamic viscosity of the water and g is the gravitational acceleration. The relationship between pressure gradient and Darcy velocity is described by the hydraulic permeability tensor **k**. The second term takes into account the dependence of the density of water on, for instance, temperature. The equation of continuity follows from the Law of Mass Conservation:

$$0 = \frac{\partial(\rho_f \Phi)}{\partial t} + \nabla(\rho_f \mathbf{v}) + \rho^* W, \tag{25}$$

where W is a source or sink term and ρ^* the corresponding fluid density. The transient flow equation is then obtained by rewriting the first term on the right side (e.g., de Marsily, 1986) and substituting for **v**:

$$\rho_f(\alpha + \beta\Phi)\frac{\partial P}{\partial t} = \nabla\left(\frac{\rho_f \mathbf{k}}{\mu}(\nabla P + \rho_f g \nabla z\right) + \rho^* W, \tag{26}$$

where α and β are the compressibility (i.e. the reciprocal of the elastic bulk modulus) of the rock matrix and the water. An alternative form of the flow equation can be obtained by the following substitutions:

$$\rho_f = \rho_0(1 + \frac{\rho_f - \rho_0}{\rho_0}) = \rho_0(1 + \rho_r) \quad \text{and} \quad h_0 = z + \frac{P}{\rho_0 g}, \tag{27}$$

where h_0 is a constant density hydraulic head at a temperature T_0 (i.e., $\rho(T_0) = \rho_0$). With

$$\nabla P = \rho_0 g(\nabla h_0 - \nabla z) \quad \text{and} \quad \frac{\partial P}{\partial t} = \rho_0 g \frac{\partial h_0}{\partial t}, \tag{28}$$

the Darcy equation (24) becomes

$$\mathbf{v} = \frac{-\rho_f g \mathbf{k}}{\mu}(\nabla h_0 + \rho_r \nabla z), \tag{29}$$

and equation (26) for fluid flow becomes

$$S_s \frac{\partial h_0}{\partial t} = \nabla\left(\frac{\rho_f g\, k}{\mu}(\nabla h_0 + \rho_r \nabla z)\right) + \frac{\rho^*}{\rho_0} W \tag{30}$$

where $S_s = \rho_f\, g\, (\alpha + \beta\Phi)$ is the specific storage coefficient. In this form, the terms on the right side of the equation correspond to the different driving forces for fluid flow, in this text referred to as 'convection' (from left to right): forced convection due to hydraulic gradients; free convection resulting from differences in density; and sources/sinks. The equation for heat transport in a porous medium, analogous to the equation for fluid flow, follows from the Law of Conservation of Energy. It is obtained from the thermal energy balance in a unit volume:

$$\frac{\partial \rho c T}{\partial t} = \nabla\big(\boldsymbol{\lambda} \nabla T - (\rho c)_f T \mathbf{v}\big) + H, \tag{31}$$

where ρc and $(\rho c)_f$ are the thermal capacity of the saturated, porous medium and the water. The individual terms on the right side of equation (31) can also be correlated with the mechanisms of heat transport (from left to right): diffusion resulting from a temperature gradient (corresponding to conduction in the steady state), advection of heat in a flow field, and sources and sinks.

The Darcy equation (eq. 29), the equation for fluid flow (eq. 30), and the equation for heat transport (eq. 31) are the basis for numerical modeling of fluid flow and heat transport discussed in chapters 3 and 4. To fully describe the interdependence of these equations, equations of state are required for the rock and fluid properties as a function of pressure and temperature. A description of the program codes used in this study, which use different equations of state, is given by Clauser (1988a) and by Kohl et al. (1993).

Index

A

B

C

D

E

F

G

H

I

J

K

L

M

N

O

P

R

S

T

U

V

Springer and the environment

Lecture Notes in Earth Sciences

For information about Vols. 1–19
please contact your bookseller or Springer-Verlag

Vol. 57: E. Lallier-Vergès, N.-P. Tribovillard, P. Bertrand (Eds.), Organic Matter Accumulation. VIII, 187 pages. 1995.

Vol. 58: G. Sarwar, G. M. Friedman, Post-Devonian Sediment Cover over New York State. VIII, 113 pages. 1995.

Vol. 59: A. C. Kibblewhite, C. Y. Wu, Wave Interactions As a Seismo-acoustic Source. XIX, 313 pages. 1996.

Vol. 60: A. Kleusberg, P. J. G. Teunissen (Eds.), GPS for Geodesy. VII, 407 pages. 1996.

Vol. 61: M. Breunig, Integration of Spatial Information for Geo-Information Systems. XI, 171 pages. 1996.

Vol. 62: H. V. Lyatsky, Continental-Crust Structures on the Continental Margin of Western North America. XIX, 352 pages. 1996.

Vol. 63: B. H. Jacobsen, K. Mosegaard, P. Sibani (Eds.), Inverse Methods. XVI, 341 pages, 1996.

Vol. 64: A. Armanini, M. Michiue (Eds.), Recent Developments on Debris Flows. X, 226 pages. 1997.

Vol. 65: F. Sansò, R. Rummel (Eds.), Geodetic Boundary Value Problems in View of the One Centimeter Geoid. XIX, 592 pages. 1997.

Vol. 66: H. Wilhelm, W. Zürn, H.-G. Wenzel (Eds.), Tidal Phenomena. VII, 398 pages. 1997.

Vol. 67: S. L. Webb, Silicate Melts. VIII. 74 pages. 1997.

Vol. 68: P. Stille, G. Shields, Radiogenetic Isotope Geochemistry of Sedimentary and Aquatic Systems. XI, 217 pages. 1997.

Vol. 69: S. P. Singal (Ed.), Acoustic Remote Sensing Applications. XIII, 585 pages. 1997.

Vol. 70: R. H. Charlier, C. P. De Meyer, Coastal Erosion – Response and Management. XVI, 343 pages. 1998.

Vol. 71: T. M. Will, Phase Equilibria in Metamorphic Rocks. XIV, 315 pages. 1998.

Vol. 72: J. C. Wasserman, E. V. Silva-Filho, R. Villas-Boas (Eds.), Environmental Geochemistry in the Tropics. XIV, 305 pages. 1998.

Vol. 73: Z. Martinec, Boundary-Value Problems for Gravimetric Determination of a Precise Geoid. XII, 223 pages. 1998.

Vol. 74: M. Beniston, J. L. Innes (Eds.), The Impacts of Climate Variability on Forests. XIV, 329 pages. 1998.

Vol. 75: H. Westphal, Carbonate Platform Slopes – A Record of Changing Conditions. XI, 197 pages. 1998.

Vol. 76: J. Trappe, Phanerozoic Phosphorite Depositional Systems. XII, 316 pages. 1998.

Vol. 77: C. Goltz, Fractal and Chaotic Properties of Earthquakes. XIII, 178 pages. 1998.

Vol. 78: S. Hergarten, H. J. Neugebauer (Eds.), Process Modelling and Landform Evolution. X, 305 pages. 1999.

Vol. 79: G. H. Dutton, A Hierarchical Coordinate System for Geoprocessing and Cartography. XVIII, 231 pages. 1999.

Vol. 80: S. A. Shapiro, P. Hubral, Elastic Waves in Random Media. XIV, 191 pages. 1999.

Vol. 81: Y. Song, G. Müller, Sediment-Water Interactions in Anoxic Freshwater Sediments. VI, 111 pages. 1999.

Vol. 82: T. M. Løseth, Submarine Massflow Sedimentation. IX, 156 pages. 1999.

Vol. 83: K. K. Roy, S. K. Verma, K. Mallick (Eds.), Deep Electromagnetic Exploration. X, 652 pages. 1999.

Vol. 84: H. V. Lyatsky, G. M. Friedman, V. B. Lyatsky. Principles of Practical Tectonic Analysis of Cratonic Regions. XX, 369 pages. 1999.

Vol. 85: C. Clauser, Thermal Signatures of Heat Transfer Processes in the Earth's Crust. X, 111 pages. 1999.